SpringerBriefs in Statistics

JSS Research Series in Statistics

The current research of statistics in Japan has expanded in several directions in line with recent trends in academic activities in the area of statistics and statistical sciences over the globe. The core of these research activities in statistics in Japan has been the Japan Statistical Society (JSS). This society, the oldest and largest academic organization for statistics in Japan, was founded in 1931 by a handful of pioneer statisticians and economists and now has a history of about 90 years. Many distinguished scholars have been members, including the influential statistician Hirotugu Akaike, who was a past president of JSS, and the notable mathematician Kiyosi Itô, who was an earlier member of the Institute of Statistical Mathematics (ISM), which has been a closely related organization since the establishment of ISM. The society has two academic journals: the *Japanese Journal of Statistics and Data Science* (JJSD, Springer), which is the successor of the *Journal of the Japan Statistical Society* (JJSS) and the *Journal of the Japan Statistical Society* (Japanese Series). The membership of JSS consists of researchers, teachers, and professional statisticians in many different fields including mathematics, statistics, engineering, medical sciences, government statistics, economics, business, psychology, education, and many other natural, biological, and social sciences. The JSS Series of Statistics aims to publish recent results of current research activities in the areas of statistics and statistical sciences in Japan that otherwise would not be available in English; they are complementary to the two JSS academic journals, both English and Japanese. Because the scope of a research paper in academic journals inevitably has become narrowly focused and condensed in recent years, this series is intended to fill the gap between academic research activities and the form of a single academic paper. The series will be of great interest to a wide audience of researchers, teachers, professional statisticians, and graduate students in many countries who are interested in statistics and statistical sciences, in statistical theory, and in various areas of statistical applications.

Masanobu Taniguchi · Diane Pierret ·
Martin Schumann · Thomas A. Severini ·
Gautam Tripathi · Yujie Xue

Econometrics, Finance, and Time Series Analysis

Masanobu Taniguchi
Department of Applied Mathematics
Waseda University
Shinjuku-ku, Tokyo, Japan

Diane Pierret
Department of Finance
University of Luxembourg
Luxembourg, Luxembourg

Martin Schumann
School of Business and Economics
Maastricht University
Maastricht, The Netherlands

Thomas A. Severini
Department of Statistics
Northwestern University
Evanston, IL, USA

Gautam Tripathi
Department of Economics and Management
University of Luxembourg
Luxembourg, Luxembourg

Yujie Xue
The Institute of Statistical Mathematics
Waseda University
Tachikawa-shi, Tokyo, Japan

ISSN 2191-544X ISSN 2191-5458 (electronic)
SpringerBriefs in Statistics
ISSN 2364-0057 ISSN 2364-0065 (electronic)
JSS Research Series in Statistics
ISBN 978-981-95-8044-6 ISBN 978-981-95-8045-3 (eBook)
https://doi.org/10.1007/978-981-95-8045-3

This Springer imprint is published by the registered company Springer Nature Singapore Pte Ltd.
The registered company address is: 152 Beach Road, #21-01/04 Gateway East, Singapore 189721, Singapore

To our families

Preface

This book presents a modern perspective on time series and panel data methods in econometrics and finance. It introduces a very general divergence measure between spectral densities and develops associated inference procedures that are both efficient and robust, thereby opening new directions for methodological research. The volume also proposes a novel measure of systemic risk in energy markets that quantifies the economic costs of energy asset distress during crisis periods, and examines the dynamic interaction between solvency and funding liquidity risk in the banking sector using panel vector autoregressive models. In addition, the book develops a new integrated likelihood approach for estimating nonlinear panel data models. Unlike existing integrated likelihood methods, the proposed approach yields a likelihood that more closely approximates a genuine parametric likelihood. The book further explains how this improvement is driven by first-order information unbiasedness, and why this property plays a more central role for inference than for point estimation. Together, the contributions in this volume illustrate recent advances in econometric methodology and their relevance for empirical research in economics and finance.

The impetus for this book arose from the 2023 interdisciplinary grant "TIME-DATA," awarded by the Institute of Advanced Study (IAS) of the University of Luxembourg to Christophe Ley (Department of Mathematics), Diane Pierret (Department of Finance), and Gautam Tripathi (Department of Economics and Management). This grant supported the 2024 visit of Masanobu Taniguchi to the University of Luxembourg as a Distinguished Professor. The authors gratefully acknowledge the generous support of the IAS and thank Professor Ley for his initiative in making Taniguchi's visit possible. Masanobu Taniguchi's research was supported in part by JSPS Grant (S) No. 18H05290.

The authors also wish to thank Professor Kunitomo, Editor of *Springer Briefs*, for his support and for facilitating the publication of this volume.

Tokyo, Japan	Masanobu Taniguchi
Luxembourg, Luxembourg	Diane Pierret
Maastricht, The Netherlands	Martin Schumann
Evanston, IL, USA	Thomas A. Severini
Luxembourg, Luxembourg	Gautam Tripathi
Tokyo, Japan	Yujie Xue

February 2026

Contents

Chapter 1
Introduction

The fields of econometrics, finance, and time series analysis continue to evolve in response to increasingly complex data structures and substantive economic questions. Empirical researchers are routinely confronted with persistent dependence over time, cross-sectional heterogeneity, nonlinear dynamics, and high-dimensional environments, all of which challenge classical methods of estimation and inference. This volume brings together a collection of chapters that address these challenges from complementary perspectives, combining theoretical advances with methodological insights and practical implications for empirical work.

The contributions span a broad range of topics in econometrics, finance, and time series analysis, reflecting the diverse tools required to study dynamic economic and financial phenomena. Some chapters revisit foundational tools—such as likelihood-based methods, spectral analysis, and shrinkage techniques—and adapt them to settings involving nonstandard asymptotics, dependent observations, or complex nuisance structures. Others demonstrate how these methodological innovations can be deployed to study economically relevant phenomena in finance and macroeconomics, including volatility transmission, systemic risk, and financial intermediation.

The volume also highlights the close interplay between theory and practice. On the methodological side, contributors propose new estimators, refine asymptotic arguments, and clarify the role of bias, efficiency, and robustness in finite samples. On the applied side, these tools are embedded in empirically relevant models and illustrated using real-world data, thereby showing how careful modeling of dependence can lead not only to improved statistical properties but also to deeper economic insight.

Another unifying feature of the book is its emphasis on temporal dependence—whether arising in regression or panel settings—and on the consequences this dependence has for estimation and inference. The chapters show how classical asymptotic arguments must often be refined or reinterpreted in settings involving short panels, persistent time series, nonlinear models, complex nuisance parameters, or correlated residual structures. At the same time, the volume illustrates how likelihood-based

M. Taniguchi et al., *Econometrics, Finance, and Time Series Analysis*,
JSS Research Series in Statistics,
https://doi.org/10.1007/978-981-95-8045-3_1

and pseudolikelihood-based approaches can be adapted to address these issues in a coherent and transparent way.

Taken together, the chapters offer a snapshot of current research at the intersection of econometrics, finance, and time series analysis, and point to directions for future work in these rapidly evolving areas. Although the individual chapters described below differ in emphasis and scope, they share a common objective: to deepen understanding of the statistical foundations underlying econometric and financial models, and to provide researchers with tools that are both theoretically well grounded and practically useful. The book is intended for researchers and advanced graduate students in econometrics, finance, and related fields, as well as for applied economists seeking insight into the methodological choices underlying contemporary empirical work.

Chapter 2 develops a frequency-domain framework for inference in Gaussian stationary time series based on the Hellinger distance between spectral densities. The chapter introduces a time series version of the Hellinger distance and uses it to study local properties of parametric spectral models, including situations where the spectral density is non-regular and standard asymptotic theory no longer applies. For regular spectral models, an estimation procedure based on minimizing the Hellinger distance between a parametric spectrum and a nonparametric spectral estimator is considered. The resulting estimator is shown to be asymptotically efficient and to provide a robust alternative to the classical Whittle estimator. Brief numerical studies are included to illustrate the finite-sample behavior of the proposed methods.

Chapter 3 develops a general framework for spectral density estimation based on a local Whittle likelihood approach. The chapter reviews both parametric and nonparametric methods for spectral estimation, highlighting the limitations of fully parametric specifications when the underlying data-generating process is unknown. To address this issue, a local Whittle likelihood based on a general score function is proposed, encompassing several existing approaches as special cases and thereby allowing for broader applicability. The chapter establishes the effective asymptotic properties of the proposed estimator and provides a detailed comparison with alternative spectral density estimators. For a particular specification of the score function, a one-step-ahead predictor is constructed, and its theoretical and numerical performance is examined. The results demonstrate that the proposed predictor achieves smaller prediction errors than the classical exponentially weighted linear predictor.

Chapter 4 examines how extreme energy price shocks affect the broader economy by estimating the exposure of the non-energy sector to adverse movements in energy prices. A vector-error correction model, combined with a multiplicative GARCH structure, is used to capture both long-run relationships and volatility spillovers among energy assets. Dynamic principal component analysis is applied to extract latent market factors, and tail expectations are estimated to construct conditional Marginal Expected Shortfall measures. These are then combined with consumption exposure data to quantify the potential losses associated with different energy assets during a price shock. The approach is applied to European energy futures and industrial equity returns, illustrating how energy market disruptions are transmitted to the broader economy.

Chapter 5 analyzes the interaction between solvency and liquidity in banking using a panel vector autoregressive specification with fixed-effects. Solvency is measured through capital-based indicators reflecting a bank's capacity to absorb losses, while liquidity risk refers to the risk that a bank cannot meet its short-term obligations without incurring substantial costs. In the empirical implementation, liquidity is proxied by the composition of short-term funding and the availability of liquid assets on the balance sheet. The model jointly estimates the dynamics of capital ratios, funding structures, and stressed solvency measures for a large panel of banks. Based on supervisory balance sheet data for U.S. bank holding companies, the analysis quantifies the feedback loops between solvency and liquidity risks that amplify financial stress. Impulse response functions show how liquidity shocks propagate to solvency, and vice versa, providing a reduced-form representation of the solvency–liquidity nexus relevant for stress testing and policy design.

Chapter 6 introduces a new integrated likelihood approach for estimating nonlinear panel data models with fixed-effects. The proposed methodology yields a likelihood that more closely approximates a genuine parametric likelihood than existing approaches. Although the supporting statistical theory is developed under asymptotics in which both the cross-sectional and time dimensions increase, simulation evidence indicates that the method performs well even in short panels.

Chapter 7 examines the role of information bias in likelihood-based estimation and inference for panel data models. It shows that improvements in small-sample performance can be attributed to first-order information unbiasedness, a property that is particularly important for inference. These insights help explain several well-established simulation findings in the panel data literature, including the excellent coverage properties of conditional likelihood-based confidence regions in short panels and the superior performance of pseudolikelihoods that are both score and information unbiased.

Chapter 8 investigates shrinkage estimation for linear regression models with dependent residuals. While the least squares estimator is a natural benchmark in linear regression, it is well known that when the dimension of the regression coefficient vector exceeds one and the residuals exhibit dependence, estimators that incorporate the covariance structure of the residual process can achieve superior mean square error performance. In this context, the chapter focuses on shrinkage estimation based on the best linear unbiased estimator, which explicitly exploits the covariance matrix of the residuals. The chapter proposes a new shrinkage estimator constructed from the best linear unbiased estimator and establishes sufficient conditions under which it improves upon the benchmark estimator in terms of mean square error. Since the covariance matrix of the residual process is typically unknown in practice, a feasible version of the proposed shrinkage estimator is developed by introducing a parametric structure for the covariance matrix and replacing unknown parameters with suitable estimates. Sufficient conditions ensuring that the feasible shrinkage estimator outperforms the benchmark estimator are also provided.

Chapter 2
Hellinger Distance Estimation for Non-regular Spectra

Abstract For Gaussian stationary process, a Time series Hellinger distance $T(f, g)$ for spectra f and g is derived. Evaluating $T(f_\theta, f_{\theta+h})$ of the form $O(h^\alpha)$, we give $1/\alpha$-consistent asymptotics of the maximum likelihood estimator of θ for non-regular spectra. For regular spectra, we introduce the minimum Hellinger distance estimator $\hat{\theta} = \arg\min_\theta T(f_\theta, \hat{g}_n)$, where $\hat{g}_n$ is a nonparametric spectral density estimator. We show that $\hat{\theta}$ is asymptotically efficient and more robust than the Whittle estimator. Brief numerical studies are provided. This chapter is mainly based on Taniguchi and Xue (2024).

Keywords Gaussian stationary process · Spectral density · Time series Hellinger distance · Non-regular spectra · Whittle estimator · Hellinger estimator · Robustness

2.1 Introduction

For independent observations, the Hellinger distance is a very important divergence which is defined by $H(p, q) = \int |p^{1/2}(x) - q^{1/2}(x)|^2 \, dx$ where p and q are the probability density functions.

For the case of regular independent observations, whose probability densities are smooth, in Beran (1977), the minimum Hellinger distance (MHD) estimator by $\hat{\theta}_n \equiv \arg\min_{\theta\in\Theta} H(p_\theta, \hat{p}_n)$ was introduced, where $\hat{p}_n$ is a nonparametric probability density estimator with sample size n, and $\{p_\theta : \theta \in \Theta\}$ is a family of parametric models. Then he showed that $\hat{\theta}_n$ is asymptotically efficient and robust for perturbation of the true probability distribution.

For independent non-regular case when p and q have discontinuity points, etc., in Ibragimov and Has' Minskii (1981) the asymptotic estimation theory based on $H(p, q)$ was developed. For parametric model $p = p_\theta$, $\theta \in \Theta$, evaluating $H(p_\theta, p_{\theta+h})$ with respect to h as $O(h^\alpha)$, they elucidated the $1/\alpha$- consistency asymptotics of the maximum likelihood estimator (MLE) for non-regular cases.

M. Taniguchi et al., *Econometrics, Finance, and Time Series Analysis*,
JSS Research Series in Statistics,
https://doi.org/10.1007/978-981-95-8045-3_2

For the analysis of count data, it was shown that the MHD estimator discounts anomalous data points in a smooth manner, effectively rejecting highly improbable counts, and that the estimator enjoys the theoretical advantage of first-order efficiency for a correctly specified model in Simpson (1987). In Lindsay (1994), an understanding of how the MHD estimator and its relatives balance these two ideals, i.e., efficiency and robustness, and developed new approaches to efficiency and robustness trade-offs.

For the problems of hypothesis testing, Ermakov (2004) derived a natural representation for the lower bound of type I and type II error probabilities by use of the Hellinger distance.

For the case of dependent observations, Taniguchi (1979) and Taniguchi (1981) introduced two divergences $D_1(f_\theta, g) = \int_{-\pi}^{\pi}\{\log f_\theta(\lambda) + g(\lambda)/f_\theta(\lambda)\}\,d\lambda$ and $D_2(f_\theta, g) = \int_{-\pi}^{\pi}\{\Phi(f_\theta(\lambda))^2 - 2\Phi(f_\theta(\lambda))\Phi(g(\lambda))\}\,d\lambda$, where $g = g(\lambda)$ is the true spectral density of a stationary process, $f_\theta = f_\theta(\lambda)$ is a fitted parametric spectral density, and $\Phi(x)$ is a bijective function. Let $I_n = I_n(\lambda)$ be the periodogram of n-observations, and let $\hat{\theta}_1 \equiv \arg\min_{\theta\in\Theta} D_1(f_\theta, I_n)$ and $\hat{\theta}_2 \equiv \arg\min_{\theta\in\Theta} D_2(f_\theta, \psi(I_n))$, where ψ is a sort of Laplace inverse transformation. Here it should be noted that, if $\Phi(x) = x$, then $\int_{-\pi}^{\pi} \Phi(I_n(\lambda))\,d\lambda$ is a consistent estimator of $\int_{-\pi}^{\pi} \Phi(g(\lambda))\,d\lambda$. But, if Φ is not a linear function (i.e., $\Phi(x) \not\equiv x$), $\int_{-\pi}^{\pi} \Phi(I_n(\lambda))\,d\lambda$ is not consistent for $\int_{-\pi}^{\pi} \Phi(g(\lambda))\,d\lambda$. Hence we need the transformation $\psi(I_n)$. Assuming the regularity conditions, Taniguchi (1979, 1981) showed that $\hat{\theta}_1$ is asymptotically efficient, and $\hat{\theta}_2$ has robustness. Furthermore, Taniguchi (1987) introduced a generalized criterion

$D_3(f_\theta, g) = \int_{-\pi}^{\pi} K\{f_\theta(\lambda)/g(\lambda)\}\,d\lambda$ where $K(x)$ is a smooth function with a unique minimum at $x = 1$. Letting $\hat{g}_n(\lambda)$ be a nonparametric spectral density estimator of $g(\lambda)$, Taniguchi (1987) showed that $\hat{\theta}_3 \equiv \arg\min_{\theta\in\Theta} D_3(f_\theta, g_n)$ is asymptotically efficient.

For a time-varying long-memory process generated by an i.i.d. innovation process, Amimour et al. (2020) derived the asymptotics of the minimum Hellinger estimators based on the Hellinger distance of probability densities of the innovation process.

In this chapter, we develop the asymptotic estimation theory for dependent observations with a new type of Hellinger distance. For Gaussian stationary process, we derive a time series Hellinger distance

$$T(f, g) = \int_{-\pi}^{\pi} \log\{\frac{1}{2}\sqrt{\frac{f(\lambda)}{g(\lambda)}} + \frac{1}{2}\sqrt{\frac{g(\lambda)}{f(\lambda)}}\}\,d\lambda. \tag{2.1}$$

Especially, we assume f^{-1} and g^{-1} are integrable. In such case, Lebesgue measures of $\{\lambda : f(\lambda) = 0\}$, $\{\lambda : g(\lambda) = 0\}$ are zero. Thus $T(f, g)$ makes sense when f and g are non-regular spectra. Here the non-regular spectra we consider are piecewise continuous spectra and spectra taking 0 at some points. Correspondingly, regular spectra mean the continuous and positive spectra. It should be noted $T(f, g)$ is different from spectral Hellinger distance

$$H_s(f, g) = \int_{-\pi}^{\pi} \{f(\lambda)^{1/2} - g(\lambda)^{1/2}\}^2\,d\lambda.$$

For non-regular spectra, evaluating $T(f_\theta, f_{\theta+h})$ with respect to h, i.e., $T(f_\theta, f_{\theta+h}) = O(h^\alpha)$, we elucidate the $1/\alpha$-consistent asymptotics of MLE like as Ibragimov and Has' Minskii (1981).

For regular spectra, fitting parametric spectral density f_θ, we propose the time series Hellinger divergence estimator $\theta^T(\hat{g}_n) = \arg\min_{\theta\in\Theta} T(f_\theta, \hat{g}_n)$ where $\hat{g}_n$ is a nonparametric spectral density estimator. Then it is shown that $\theta^T(\hat{g}_n)$ is asymptotically efficient. Further, for a small perturbation $g(\lambda) = f_\theta(\lambda) + \varepsilon h(\lambda)$, $\varepsilon > 0$, we show that $\theta^T(\hat{g}_n)$ is more robust than Whittle type estimator, which is an approximated MLE.

This chapter is organized as follows. Section 2.2 derives the time series Hellinger distance $T(f, g)$ for non-regular spectra. Evaluating $T(f_\theta, f_{\theta+h}) = O(n^\alpha)$, we elucidate the $1/\alpha$-asymptotics of the MLE in Sect. 2.3. Section 2.4 introduces the Hellinger distance estimator $\hat{\theta} = \arg\min_\theta T(f_\theta, \hat{g}_n)$ for regular spectra. Then we show that $\hat{\theta}$ is asymptotically efficient and that $\hat{\theta}$ is more robust than the Whittle estimator for small perturbation. Small numerical studies will be given to confirm the robustness of $\hat{\theta}$.

2.2 Approximation of the Hellinger Distance Between Non-regular Spectra

In this section we discuss the spectral analysis of stationary processes. For the fundamental concepts, the readers can find the definitions and explanations of stationary process, spectral analysis, and spectral estimation in, e.g., Koopmans (1974) (p. 78, p. 74–78, and Chap. 9, respectively). Suppose that $\{X_t\}$ is a Gaussian stationary process with zero mean and spectral density $f(\lambda) \in$, where is the class of piecewise continuous spectral densities on $[-\pi, \pi] - \{0\}$, satisfying $\int_{-\pi}^{\pi} \log f(\lambda)\,\mathrm{d}\lambda > -\infty$ to avoid the singular case of a deterministic process (see Anderson 1971, p. 422). Let $\boldsymbol{X}_n = (X_1, X_2, \ldots, X_n)'$, then the likelihood function under $f \in$ is given by

$$L_n(f) = \frac{1}{(2\pi)^{n/2}} \frac{1}{|V_n(f)|^{1/2}} \exp[-\frac{1}{2}\boldsymbol{X}_n' V_n(f)^{-1}\boldsymbol{X}_n], \tag{2.2}$$

where $V_n(f) = \{\int_{-\pi}^{\pi} \mathrm{e}^{\mathrm{i}(t-s)\lambda} f(\lambda)\,\mathrm{d}\lambda;\, t, s = 1, 2, \ldots, n\}$. For another $g \in$, the likelihood ratio is

$$LR_n(f : g) \equiv \frac{L_n(g)}{L_n(f)}. \tag{2.3}$$

Introduce

$$A_n(f : g) \equiv \mathrm{E}_f[\sqrt{LR_n(f : g)}], \tag{2.4}$$

which is equivalent to Hellinger distance when deriving the asymptotics of MLE in the case of independent observations. Then we have

Theorem 2.1

$$\lim_{n\to\infty}\frac{1}{n}\log A_n(f:g) = -\frac{1}{4\pi}T(f,g), \tag{2.5}$$

where

$$T(f,g)=\int_{-\pi}^{\pi}\log\{\frac{1}{2}\sqrt{\frac{f(\lambda)}{g(\lambda)}}+\frac{1}{2}\sqrt{\frac{g(\lambda)}{f(\lambda)}}\}\,d\lambda, \tag{2.6}$$

and we call $T(f,g)$ as a time series Hellinger distance.

Proof. We prove the following four cases:

(i) f and g are continuous on $[-\pi,\pi]$ and $f, g>0$ on $[-\pi,\pi]$;
(ii) f and g are piecewise continuous on $[-\pi,\pi]$ & $f, g > 0$;
(iii) f and g are long-memory spectra;
(iv) f and g are the spectra of MA unit root processes.

Case (i) Suppose that f and g are continuous and positive on $[-\pi,\pi]$. We observe that

$$\begin{aligned}\sqrt{LR_n(f:g)} &= \exp\cdot\log\sqrt{\frac{\frac{1}{|V_n(g)|^{1/2}}\exp[-\frac{1}{2}\boldsymbol{x}_n' V_n(g)^{-1}\boldsymbol{x}_n]}{\frac{1}{|V_n(f)|^{1/2}}\exp[-\frac{1}{2}\boldsymbol{x}_n' V_n(f)^{-1}\boldsymbol{x}_n]}}\\ &= \exp[-\frac{1}{4}\log|V_n(g)|/|V_n(f)| - \frac{1}{4}\boldsymbol{x}_n'(V_n(g)^{-1}-V_n(f)^{-1})\boldsymbol{x}_n].\end{aligned} \tag{2.7}$$

Hence we obtain

$$\begin{aligned}\mathrm{E}_f\sqrt{LR_n(f:g)} &= |V_n(g)|^{-1/4}/|V_n(f)|^{-1/4}\int\cdots\int\exp[-\frac{1}{4}\boldsymbol{x}_n'(V_n(g)^{-1}\\ &\quad - V_n(f)^{-1})\boldsymbol{x}_n]\\ &\quad\times\frac{1}{(2\pi)^{n/2}}\frac{1}{|V_n(f)|^{1/2}}\exp[-\frac{1}{2}\boldsymbol{x}_n' V_n(f)^{-1}\boldsymbol{x}_n]\,d\boldsymbol{x}_n\\ &= (2\pi)^{-\frac{n}{2}}|V_n(g)|^{-1/4}|V_n(f)|^{-1/4}\\ &\quad\times\int\cdots\int\exp[-\frac{1}{4}\boldsymbol{x}_n'(V_n(g)^{-1}+V_n(f)^{-1})\boldsymbol{x}_n\,d\boldsymbol{x}_n\\ &= |V_n(g)|^{-1/4}|V_n(f)|^{-1/4}|\frac{1}{2}V_n(g)^{-1}+\frac{1}{2}V_n(f)^{-1}|^{-1/2}\\ &= |V_n(g)^{-1/2}|^{-1/2}|\frac{1}{2}V_n(g)+\frac{1}{2}V_n(f)|^{-1/2}|V_n(f)^{-1/2}|^{-1/2}.\end{aligned} \tag{2.8}$$

It is seen that

$$
\begin{aligned}
\frac{1}{n}\log \mathrm{E}_f\sqrt{LR_n(f:g)} &= -\frac{1}{2n}\left[-\frac{1}{2}\log|V_n(g)| + \log|\frac{1}{2}V_n(g) + \frac{1}{2}V_n(f)| - \frac{1}{2}\log|V_n(f)|\right] \\
&\xrightarrow{n\to\infty} -\frac{1}{2}[\frac{1}{2\pi}\int_{-\pi}^{\pi} -\frac{1}{2}\log g(\lambda)\,\mathrm{d}\lambda + \frac{1}{2\pi}\int_{-\pi}^{\pi}\log\{\frac{1}{2}g(\lambda) \\
&\quad + \frac{1}{2}f(\lambda)\}\,\mathrm{d}\lambda + \frac{1}{2\pi}\int_{-\pi}^{\pi} -\frac{1}{2}\log g(\lambda)\,\mathrm{d}\lambda] \\
&\qquad \text{(by Hannan (1970) p.396)} \\
&= -\frac{1}{2}\left[\frac{1}{2\pi}\int_{-\pi}^{\pi}\left\langle \log g(\lambda)^{-1/2} + \log\{\frac{1}{2}g(\lambda) + \frac{1}{2}f(\lambda)\} + \log f(\lambda)^{-1/2}\right\rangle \mathrm{d}\lambda\right] \\
&= -\frac{1}{2}\frac{1}{2\pi}\int_{-\pi}^{\pi}\log\left[\frac{1}{2}\sqrt{\frac{f(\lambda)}{g(\lambda)}} + \frac{1}{2}\sqrt{\frac{g(\lambda)}{f(\lambda)}}\}\right]\mathrm{d}\lambda.
\end{aligned}
$$

Case (ii) Let $_1$ be the class of the spectral densities which have the following form:

$$
f(\lambda) = \begin{cases} f_1(\lambda), & \lambda \in [0, \tau_1) \\ f_2(\lambda), & \lambda \in [\tau_1, \tau_2) \\ \vdots & \vdots \\ f_p(\lambda), & \lambda \in [\tau_{p-1}, \pi] \end{cases} \qquad (0 < \tau_1 < \cdots < \tau_{p-1} < \pi)
$$

where $f_k(\lambda)'s$ are positive continuous, and $f_k(\tau_k^-) \neq f_{k+1}(\tau_k^+)$, and $f(\lambda) = f(-\lambda)$ for $\lambda \in [-\pi, 0]$. If f and $g \in$ $_1$, we can construct smooth versions $\tilde{f}_\delta$ and $\tilde{g}_\delta$ of f and g so that

$$
\tilde{f}_\delta(\lambda) \to f(\lambda), \text{ and } \tilde{g}_\delta(\lambda) \to g(\lambda) \text{ on } [-\pi, \pi], \text{ as } \delta \searrow 0 \tag{2.9}
$$

(see Billingsley 1968, p. 41 or Taniguchi 2008, p. 157). For $\tilde{f}_\delta$ and $\tilde{g}_\delta$, (2.5) holds. Also it is shown that

$$
\log\det V_n(f) - \log\det V_n(\tilde{f}_\delta) = O(n) \times O(\delta). \tag{2.10}
$$

From this, for f and $g \in$ $_1$, (2.5) holds.

Case (iii) Let $_2$ be the class of long-memory spectral densities of the form $f(\lambda) = \frac{1}{2\pi}\frac{1}{|1-\mathrm{e}^{\mathrm{i}\lambda}|^{2d}} f_0(\lambda)$, $(0 < d < 1/2)$, $f_0(\lambda) \in$. If f and $g \in$ $_2$, as in the case of (ii), we can construct smooth versions $\tilde{f}_\delta(\lambda)$ and $\tilde{g}_\delta(\lambda)$ of $f(\lambda)$ and $g(\lambda)$, respectively, such that $\tilde{f}_\delta(\lambda) \to f(\lambda)$, and $\tilde{g}_\delta(\lambda) \to g(\lambda)$, as $\delta \searrow 0$. For $\tilde{f}_\delta$ and $\tilde{g}_\delta$, (2.5) holds. Hence, similarly as in Case (ii), for f and $g \in$ $_2$, we can show (2.5).

Case (iv) Let $_3$ be the class of spectral densities of the form $f(\lambda) = |(\mathrm{e}^{\mathrm{i}\lambda} - z_1)\cdots(\mathrm{e}^{\mathrm{i}\lambda} - z_q)|^2 f_0(\lambda)$, where $z_j = \mathrm{e}^{\mathrm{i}\lambda_j}$ ($\lambda_1, \ldots, \lambda_q$ are unequal to each other), and $f_0 \in$. If f and $g \in$ $_3$, from Dzhaparidze (1986), p. 57, we can see

$$\lim_{n\to\infty} \frac{1}{n} A_n(f : g) = -\frac{1}{4\pi} T(f_0, g_0). \tag{2.11}$$

If f and g are of the forms: $f(\lambda) = \frac{1}{2\pi}|1 - (1 - h_1)e^{i\lambda}|^2$, $g(\lambda) = \frac{1}{2\pi}|1 - (1 - h_2)e^{i\lambda}|^2$, $0 \leq h_1 \neq h_2 < 1$, similarly as in (ii), we can take the smooth versions $\tilde{f_\delta}$ and $\tilde{g}_\delta$ of f and g satisfying $\tilde{f_\delta}(\lambda) \to f(\lambda)$, and $\tilde{g}_\delta(\lambda) \to g(\lambda)$, as $\delta \searrow 0$. Then we get (2.5). □

2.3 Asymptotics of MLE in View of the Hellinger Distance

Let $\{X_t\}$ be a Gaussian stationary process with spectral density $f_\theta(\lambda) \in$, $\theta \in \Theta$ (open set of $\mathbf{R}^1$). In this section, we describe the asymptotics of MLE for θ in view of the time series Hellinger distance $T(f_\theta, f_{\theta+h})$, where $\theta + h \in \Theta$. Assume that $f_\theta(\lambda)$ belongs to , and f_θ is piecewisely two times continuously differentiable with respect to $\theta \in \Theta$ for almost all $\lambda \in [-\pi, \pi]$.

Theorem 2.2 *Suppose that*

(i) there exists a number $\alpha > 0$ *such that*

$$\sup_{\theta\in\Theta} \sup_{h} |h|^{-\alpha} T(f_\theta, f_{\theta+h}) = A < \infty, \tag{2.12}$$

(ii) for any compact set $K \subset \Theta$, *there exists* $a = a(K) > 0$ *such that*

$$T(f_\theta, f_{\theta+h}) \geq \frac{a|h|^\alpha}{1 + |h|^\alpha}, \qquad \theta \in K. \tag{2.13}$$

Then the maximum likelihood estimator $\hat{\theta}_n$ *of* θ *is defined, and is consistent and*

$$\sup_{\theta\in K} P_\theta[n^{1/\alpha}|\hat{\theta}_n - \theta| > H] \leq B_0 \exp\{-b_0 a H^\alpha\}, \tag{2.14}$$

where the positive constants B_0 *and* b_0 *do not depend on* K.

Proof. From Theorem 2.1, and following the line of Theorem 5.3 of Ibragimov and Has' Minskii (1981), where the i.i.d. case is proved, it can be proved. □

In what follows we evaluate $T(f_\theta, f_{\theta+h})$ for (2.12) and (2.13).

Case (i). Suppose that f_θ is continuous and positive with respect to $\lambda \in [-\pi, \pi]$, and f_θ twice continuously differentiable with respect to θ. Then we obtain

$$
\begin{aligned}
T(f_\theta, f_{\theta+h}) &= \int_{-\pi}^{\pi} \log\{\frac{1}{2}\sqrt{\frac{f_\theta}{f_{\theta+h}}} + \frac{1}{2}\sqrt{\frac{f_{\theta+h}}{f_\theta}}\} \, \mathrm{d}\lambda \\
&= h \int_{-\pi}^{\pi} \left[\frac{1}{2}\{(-\frac{1}{2})(\frac{1}{f_\theta}\frac{\partial}{\partial\theta} f_\theta) + (\frac{1}{2})(\frac{1}{f_\theta}\frac{\partial}{\partial\theta} f_\theta)\} \right] \mathrm{d}\lambda + O(h^2) \\
&= O(h^2) \quad \text{uniformly } \theta \in \Theta,
\end{aligned}
$$

which leads to (2.12) and (2.13) with $\alpha = 2$. Hence the consistency order of the MLE $\hat{\theta}_n$ is of order $\sqrt{n}$.
Case (ii) Let $f_\theta \in {}_1$ with $p = 2$, and $\theta = \tau_1$. Then, it is not difficult to see

$$
T(f_\theta, f_{\theta+h}) = O(h) \quad \text{uniformly } \theta \in \Theta,
$$

which implies the consistent order of MLE for the jump point is of order n.

Case (iii) Let $f_\theta \in {}_2$, and let θ be the long-memory parameter d $(0 < d < 1/2)$, i.e. $\theta = d$. Then,

$$
\begin{aligned}
T(f_\theta, f_{\theta+h}) &= \int_{-\pi}^{\pi} \log\frac{1}{2}(|1 - \mathrm{e}^{\mathrm{i}\lambda}|^{-h} + |1 - \mathrm{e}^{\mathrm{i}\lambda}|^{h}) \, \mathrm{d}\lambda \\
&= \int_{-\pi}^{\pi} \left[h \log|1 - \mathrm{e}^{\mathrm{i}\lambda}| + O(h^2) \right] \mathrm{d}\lambda \\
&= O(h^2),
\end{aligned}
$$

which implies the consistent order of MLE for θ is of order $\sqrt{n}$.

Case (iv) Let $f_\theta \in G_3$ and

$$
f_\theta(\lambda) = |(\mathrm{e}^{\mathrm{i}\lambda} - z_1) \cdots (\mathrm{e}^{\mathrm{i}\lambda} - z_1)|^2 f_{0,\theta}(\lambda),
$$

where $f_{0,\theta}$ is continuously two times differentiable with respect to θ. Then, similarly as in (i), we obtain

$$
T(f_\theta, f_{\theta+h}) = O(h^2) \quad \text{uniformly } \theta \in \Theta,
$$

which implies the consistent order of MLE for d is of order $\sqrt{n}$.
In the case when f_θ has near unit root, using the relation

$$
\frac{|1 - \mathrm{e}^{\mathrm{i}\lambda}|}{|1 - (1-h)\mathrm{e}^{\mathrm{i}\lambda}|} + \frac{|1 - (1-h)\mathrm{e}^{\mathrm{i}\lambda}|}{|1 - \mathrm{e}^{\mathrm{i}\lambda}|} \doteqdot \frac{h}{|1 - \mathrm{e}^{\mathrm{i}\lambda}|} + \frac{h}{|1 - \mathrm{e}^{\mathrm{i}\lambda}|},
$$

we can see $T(f_\theta, f_{\theta+h}) = O(h)$, which implies the consistent order of MLE for θ is of order n.

2.4 Estimation Theory

In Sect. 2.2, we introduced the time series Hellinger type divergence $T(f, g)$, and in Sect. 2.3, we elucidate the asymptotics of the MLE for time series processes. It has been known that for many parametric families of distributions, the MLE has full asymptotic efficiency. But it has been recognized that MLE is not stable under small perturbations in the underlying model. In view of this, by fitting a parametric probability density f_θ to the true probability density g by the Hellinger distance (Beran 1977) has introduced an efficient estimator of θ, which is stable under perturbations. For this situation, the fitting of parametric model f_θ to the true one g is appropriate.

For estimation of time series spectra, by fitting parametric spectral density f_θ to the true spectral density g, Taniguchi (1979) and Taniguchi (1987) introduced new estimators of θ, and discussed the efficiency and robustness. To find parametric spectral family $= \{f_\theta\}$ is as follows. The above two papers assumed that θ is vector-valued. In such cases, if we assume θ_j is the jth coefficient of $AR(p)$ model, so that $\theta = (\theta_1, \dots, \theta_p)'$, then $= \{AR(p)$ spectral densities$\}$. If we take p sufficiently large, then f_θ approximates g sufficiently. So we assume is an abstract class of the parametric models, which approximates g suitably. Of course, $= \{ARMA(p, q)\}$, and $= \{MA(q)\}$ are appropriate. By the use of $T(f, g)$, this section develops the estimation theory for regular spectra of stationary processes. Let $\{X_t\}$ be a Gaussian process with zero mean and spectral density $g(\lambda)$. In our setting, we assume that θ is a scalar-valued, so that the discussion meets Ibragimov and Has'minskii's framework. But this is not restrictive. Suppose that $f_\theta = \frac{\sigma^2}{2\pi}|\sum_{j=0}^{p} a_j e^{ij\lambda}|^{-2}$ (AR(p)-spectral density), and that $a_j = a_j(\theta)$, $j = 1, \dots, p$, $\sigma^2 = \sigma^2(\theta)$, $\theta \in \Theta \subset \mathbb{R}^1$. Or if we assume that θ_j is the jth coefficient of $AR(p)$ model, then we may take our θ as $\theta = c_1\theta_1 + \cdots + c_p\theta_p$, where $c_1, \dots, c_p$ are arbitrarily known constants. Hence our $= \{f_\theta;\ \theta \in \Theta\}$ can approximate general spectra g suitablely.

In order to estimate θ, we propose:

$$T(f_\theta, g) = \int_{-\pi}^{\pi} \log\{\frac{1}{2}\sqrt{\frac{f_\theta}{g}} + \frac{1}{2}\sqrt{\frac{g}{f_\theta}}\}\, d\lambda. \tag{2.15}$$

As a benchmark, we may use the Whittle divergence:

$$W(f_\theta, g) \equiv \int_{-\pi}^{\pi} \{\log \frac{f_\theta}{g} + \frac{g}{f_\theta}\}\, d\lambda, \tag{2.16}$$

which corresponds to a Gaussian likelihood. Based on them we define the pseudo-true values by

$$\theta^T(g) = \arg\min_{\theta \in \Theta} T(f_\theta, g), \tag{2.17}$$

and

$$\theta^W(g) = \arg\min_{\theta \in \Theta} W(f_\theta, g), \tag{2.18}$$

respectively. Since $\theta^T(g)$ and $\theta^W(g)$ may be multiple-valued, we shall use these notations to indicate any one of the possible values.

Suppose that $\{X_t\}$ is a Gaussian stationary process with $\mathrm{E}(X_t) = 0$, and has the spectral density $g(\lambda)$, and the autocovariance function $R(s) = \mathrm{E}\{X_t X_{t+s}\}$.

Assumption 2.1

$$\sum_{s=-\infty}^{\infty} |s||R(s)| < \infty.$$

Let $\{X_1, X_2, \ldots, X_n\}$ be an observed stretch of $\{X_t\}$. To estimate $\theta^T(g)$ and $\theta^W(g)$, we use a nonparametric window type spectral estimator $\hat{g}_n(\lambda)$ whose window $K(x)$ satisfies,

Assumption 2.2 (i) $K(t)$ is bounded, even, non-negative and such that

$$\int_{-\infty}^{\infty} K(t)\,\mathrm{d}t = 1,$$

(ii) For $M = O(n^\alpha)$, $(1/4 < \alpha < 1/2)$, the function $K_n(\lambda) = MK(M\lambda)$ can be expanded as

$$K_n(\lambda) = \frac{1}{2\pi}\sum_l k(\frac{l}{M})\mathrm{e}^{-il\lambda},$$

where $k(x)$ is a continuous, even function with $k(0) = 1$, $|k(x)| \le 1$ and $\int_{-\infty}^{\infty} k(x)^2 < \infty$, and $\lim_{|x|\to 0} \frac{1-k(x)}{|x|^2} = \kappa < \infty$.

Henceforth we use the following nonparametric spectral estimator:

$$\hat{g}_n(\lambda) = \frac{2\pi}{n}\sum_{j=1}^{n} K_n(\lambda - \lambda_j) I_n(\lambda_j), \tag{2.19}$$

where $\lambda_j = -\pi + 2\pi j/n$. From Theorems 9 and 10 in Hannan (1970) Section V, it follows that

$$\mathrm{E}[\{\hat{g}_n(\lambda) - g(\lambda)\}^2] = O(M/n), \qquad \text{uniformly in } \lambda. \tag{2.20}$$

It should be mentioned that for the local Whittle likelihood estimator $\breve{g}_n$ discussed in Xue and Taniguchi (2020), $\breve{g}_n$ also has the similar property to (2.20). Then, Theorems 2 and 4 of Taniguchi (1987) essentially yield.

Proposition 2.1 *Suppose that $\theta^T(g)$ and $\theta^W(g)$ exist uniquely and lies in* Int(Θ). *Then it holds that, as $n \to \infty$,*

(i)

$$\theta^T(\hat{g}_n) \to \theta^T(g) \textit{ and } \theta^W(\hat{g}_n) \to \theta^W(g) \textit{ in } \quad, \tag{2.21}$$

(ii) $\sqrt{n}\{\theta^T(\hat{g}_n) - \theta^T(g)\}$ *and* $\sqrt{n}\{\theta^W(\hat{g}_n) - \theta^W(g)\}$ *are asymptotically normal,*

(iii) if $g = f_{\theta_0}$, *then* $\theta^T(g) = \theta_0$ *and* $\theta^W(g) = \theta_0$. *Also* $\sqrt{n}\{\theta^T(\hat{g}_n) - \theta_0)\}$ *and* $\sqrt{n}\{\theta^W(\hat{g}_n) - \theta_0)\}$ *converge to* $N(0, F(\theta_0)^{-1})$, *where* $F(\theta_0) = \frac{1}{4\pi}[\int_{-\pi}^{\pi} \frac{\partial}{\partial\theta} \log f_\theta(\lambda) \frac{\partial}{\partial\theta'} \log f_\theta(\lambda)\, \mathrm{d}\lambda]|_{\theta=\theta_0}$.

Hence, if $g = f_{\theta_0}$, the both estimators $\theta^T(\hat{g}_n)$ and $\theta^W(\hat{g}_n)$ are asymptotically efficient, and are asymptotically equivalent.

In what follows, setting $g(\lambda) = f_{\theta_0}(\lambda) + \varepsilon h(\lambda)$, $\varepsilon > 0$, we compare the robustness of the two estimates when there exists perturbation. Our model is a signal +(small) noise model. For example, Hannan (1970) discussed the prediction etc. By comparing at the point θ_0 which divergence is least affected by perturbation, we show which estimator approximates the real parameter θ_0 better. Write $\frac{g(\lambda)}{f_{\theta_0}(\lambda)} = 1 + \varepsilon \frac{h(\lambda)}{f_{\theta_0}(\lambda)} = 1 + \varepsilon\delta(\lambda)$ (say).

Expanding the score functions with respect to ε, we obtain

$$\log\{\frac{1}{2}\sqrt{\frac{f_{\theta_0}(\lambda)}{g(\lambda)}} + \frac{1}{2}\sqrt{\frac{g(\lambda)}{f_{\theta_0}(\lambda)}}\} = \frac{\varepsilon^2}{8}\{\delta(\lambda)\}^2 + O(\varepsilon^3), \tag{2.22}$$

and

$$\log \frac{f_{\theta_0}(\lambda)}{g(\lambda)} + \frac{g(\lambda)}{f_{\theta_0}(\lambda)} = 1 + \frac{\varepsilon^2}{2}\{\delta(\lambda)\}^2 + O(\varepsilon^3). \tag{2.23}$$

Then

$$|T(f_{\theta_0}, f_{\theta_0}) - T(f_{\theta_0}, g)| = \frac{\varepsilon^2}{8}\int_{-\pi}^{\pi} \{\delta(\lambda)\}^2\, \mathrm{d}\lambda + O(\varepsilon^3),$$

and

$$|W(f_{\theta_0}, f_{\theta_0}) - W(f_{\theta_0}, g)| = \frac{\varepsilon^2}{2}\int_{-\pi}^{\pi} \{\delta(\lambda)\}^2\, \mathrm{d}\lambda + O(\varepsilon^3),$$

which implies $T(f_\theta, g)$ is more robust than $W(f_\theta, g)$ for the perturbation of $g(\lambda)$. Our robustness is a sort of the influence curve introduced by Hampel (1974). Our formulae (2.22) and (2.23) are a sort of influence curve, which shows a sort of "sensitiveness of estimators".

Next we discuss the numerical aspect. We consider the situation where the observations of AR(1) model is contaminated by a MA(1) model. Let $\{e_t\} \sim$ i.i.d.$N(0, 1)$, $\{u_t\} \sim$ i.i.d.$N(0, 1)$, and $\{e_t\}$ and $\{u_t\}$ are mutually independent.

Introduce

$$Z_t = \theta_0 Z_{t-1} + e_t, \quad (|\theta| < 1),$$

$$W_t = u_t + \eta u_{t-1}, \quad (|\eta| < 1),$$

Table 2.1 The MSE when $\varepsilon = 0.05$

η	−0.8		−0.4		−0.2		0.2		0.4		0.8	
θ_0	T	W	T	W	T	W	T	W	T	W	T	W
−0.8	0.0282	0.0491	0.0285	0.0499	0.0282	0.0482	0.0287	0.0504	0.0289	0.0490	0.0292	0.0498
−0.6	0.0191	0.0302	0.0190	0.0298	0.0197	0.0303	0.0198	0.0306	0.0195	0.0291	0.0198	0.0306
−0.4	0.0362	0.0392	0.0346	0.0373	0.0356	0.0383	0.0355	0.0382	0.0349	0.0382	0.03642	0.0390
−0.2	0.0318	0.0324	0.0329	0.0338	0.0330	0.0339	0.0326	0.0337	0.0325	0.0329	0.0330	0.0338
0.2	0.0282	0.0315	0.0290	0.0328	0.0292	0.0330	0.0287	0.0324	0.0296	0.0335	0.0289	0.0326
0.4	0.0287	0.0298	0.0272	0.0286	0.0275	0.0290	0.0284	0.0297	0.0281	0.0290	0.0282	0.0321
0.6	0.0212	0.0469	0.0204	0.0438	0.0206	0.02450	0.0210	0.0467	0.0202	0.0460	0.0207	0.0434
0.8	0.0139	0.0383	0.0141	0.0387	0.0140	0.0388	0.0141	0.0387	0.0136	0.0385	0.0141	0.0386

Table 2.2 The MSE when $\varepsilon = 0.1$

η	−0.8		−0.4		−0.2		0.2		0.4		0.8	
θ_0	T	W	T	W	T	W	T	W	T	W	T	W
−0.8	0.0281	0.0490	0.0287	0.0503	0.0281	0.0473	0.0295	0.0508	0.0302	0.0492	0.0311	0.0513
−0.6	0.0191	0.0306	0.019	0.0310	0.0203	0.0307	0.0208	0.0331	0.0206	0.0303	0.0216	0.0298
−0.4	0.0369	0.0401	0.0340	0.0366	0.0361	0.0387	0.0363	0.0389	0.0354	0.0364	0.0391	0.0395
−0.2	0.0305	0.0310	0.0330	0.0339	0.0335	0.0344	0.0329	0.0343	0.0328	0.0330	0.0342	0.0350
0.2	0.0303	0.0275	0.0327	0.0291	0.0333	0.0295	0.0320	0.0284	0.0344	0.0303	0.0324	0.0288
0.4	0.0307	0.0314	0.027	0.0259	0.0274	0.0292	0.0288	0.0303	0.0283	0.0290	0.0284	0.0324
0.6	0.0236	0.0487	0.0211	0.0418	0.0215	0.0446	0.0219	0.0487	0.0202	0.0446	0.0218	0.0407
0.8	0.0146	0.0385	0.0148	0.0390	0.0145	0.0395	0.0144	0.0388	0.0136	0.0386	0.0145	0.0388

and define $Y_t = Z_t + \varepsilon W_t$. Then the contaminated observations can be simulated by Y_t with the spectral density

$$g(\lambda) = \frac{1}{2\pi}\frac{1}{|1-\theta_0 e^{i\lambda}|^2} + \varepsilon^2 \frac{1}{2\pi}|1+\eta e^{i\lambda}|^2.$$

For $\theta_0 = -0.8, \ldots, 0.8$, $\eta = -0.8, \ldots, 0.8$, and $\varepsilon = 0.05$ and 0.1, we generate the date $Y_1, \ldots, Y_n$. Letting $f_\theta(\lambda) = \frac{1}{2\pi}\frac{1}{|1-\theta e^{i\lambda}|^2}$, calculate $\theta^T(\hat{g}_n)$ and $\theta^W(\hat{g}_n)$, and repeat this procedure 100 times. Then we obtain the following tables. In Tables 2.1 and 2.2, T and W stands for Hellinger distance and Whittle divergence respectively. The mean square errors (MSEs) of $\theta^T(\hat{g}_n)$ and $\theta^W(\hat{g}_n)$ are shown when $\varepsilon = 0.05$ and 0.1. Obviously, the MSEs of the Hellinger distance estimator $\theta^T(\hat{g}_n)$ are smaller than those of Whittle divergence estimator, which agrees with the theoretical results.

References

Amimour, A., K. Belaide, and O. Hili (2020). "Minimum Hellinger distance estimates for a periodically time-varying long memory parameter". In: *arXiv preprint*.

Anderson, Theodore W (1971). *The statistical analysis of time series*. New York: Springer.

Beran, Rudolf (1977). "Minimum Hellinger distance estimates for parametric models". In: *The annals of Statistics* 5.3, pp. 445–463.

Billingsley, P. (1968). *Convergence of Probability Measures*. New York: John Wiley.

Dzhaparidze, K. (1986). *Parameter Estimation and Hypothesis Testing in Spectral Analysis of Stationary Time Series*. Springer Science & Business Media.

Ermakov, Michael Sergeevich (2004). "On asymptotically efficient statistical inference for moderate deviation probabilities". In: *Theory of Probability & Its Applications* 48.4, pp. 622–641.

Hampel, Frank R (1974). "The influence curve and its role in robust estimation". In: *Journal of the american statistical association* 69.346, pp. 383–393.

Hannan, E. J. (1970). *Multiple Time Series*. New York: Wiley.

Ibragimov, Ildar Abdulovich and Rafail Zalmanovich Has' Minskii (1981). *Statistical estimation: asymptotic theory*. New York: Springer.

Koopmans, Lambert H (1974). *The spectral analysis of time series*. London: Academic Press.

Lindsay, Bruce G (1994). "Efficiency versus robustness: The case for minimum Hellinger distance and related methods". In: *The annals of statistics* 22.2, pp. 1081–1114.

Simpson, Douglas G (1987). "Minimum Hellinger distance estimation for the analysis of count data". In: *Journal of the American statistical Association* 82.399, pp. 802–807.

Taniguchi, Masanobu (1979). "On estimation of parameters of Gaussian stationary processes". In: *Journal of Applied Probability* 16.3, pp. 575–591.

Taniguchi, Masanobu (1981). "An estimation procedure of parameters of a certain spectral density model". In: *Journal of the Royal Statistical Society: Series B (Methodological)* 43.1, pp. 34–40.

Taniguchi, Masanobu (1987). "Minimum contrast estimation for spectral densities of stationary processes". In: *Journal of the Royal Statistical Society: Series B (Methodological)* 49.3, pp. 315–325.

Taniguchi, Masanobu (2008). "Non-regular estimation theory for piecewise continuous spectral densities". In: *Stochastic processes and their applications* 118.2, pp. 153–170.

Taniguchi, Masanobu and Yujie Xue (2024). "Hellinger distance estimation for nonregular spectra". In: *Theory of Probability & Its Applications* 69.1, pp. 150–160.

Xue, Yujie and Masanobu Taniguchi (2020). "Local Whittle likelihood approach for generalized divergence". In: *Scandinavian Journal of Statistics* 47.1, pp. 182–195.

Chapter 3
Local Whittle Likelihood Approach for Generalized Divergence

Abstract There are many approaches in the estimation of spectral density. With regard to parametric approaches, different divergences are proposed in fitting a certain parametric family of spectral densities. Moreover, nonparametric approaches are also quite common considering the situation when we cannot specify the model of process. In this chapter we develop a local Whittle likelihood approach based on a general score function, with some special cases of which, the approach applies to more applications. This chapter highlights the effective asymptotics of our general local Whittle estimator, and presents a comparison with other estimators. Additionally, for a special case, we construct the one-step ahead predictor based on the form of the score function. Subsequently, we show that it has a smaller prediction error than the classical exponentially weighted linear predictor. The provided numerical studies show some interesting features of our local Whittle estimator. This chapter is mainly based on Xue and Taniguchi (2020).

Keywords Spectral density · local Whittle estimator · Nonparametric spectral estimator · Kernel function · Exponentially weighted predictor

3.1 Introduction

There are many approaches in the estimation of the spectral density. With regard to parametric approaches, Taniguchi (1980) proposed three divergences in fitting a certain parametric family of spectral densities $\{f_\theta(\lambda); \theta \in \Theta\}$ to a Gaussian stationary process with mean 0 and true spectral density $f(\lambda)$, in which the Whittle likelihood is included. Furthermore, some of the proposed methods were extended to multi-dimensional cases, and the divergences were generalized by Taniguchi (1987) for fourth-order stationary processes by introducing a nonparametric window type estimator $\hat{g}_N(\lambda)$. Moreover, nonparametric approaches are also quite common because we often cannot specify the model of the process. They are based on a smoothed periodogram (Hannan 1970). Additionally, for the multivariate stationary process, the approach of "moving blocks" resampling scheme can be used to estimate the

M. Taniguchi et al., *Econometrics, Finance, and Time Series Analysis*,
JSS Research Series in Statistics,
https://doi.org/10.1007/978-981-95-8045-3_3

spectral density for an m-dependent process (Liu et al., 1992) and α-mixing process (Politis and Romano 1992).

Hjort and Jones (1996) introduced a nonparametric probability density estimator with parametric overtones, called the local likelihood estimator. The local likelihood estimator can be considered as a replacement for the smoothed estimator for probability density; it was shown that the bias of the former is potentially smaller than that of the later. (Tjøstheim and Ove Hufthammer 2013) used the local likelihood estimator to approximate the local bivariate density such that the local dependence between the financial objects can be measured using a new approach.

In this study, we develop a local Whittle likelihood approach for a score function $\rho_\theta(\lambda)$. Based on $\rho_\theta(\cdot)$, we introduce a local Whittle estimator as the minimizer of the local Whittle likelihood function and denote it as $\hat{\theta} = \hat{\theta}(\lambda)$ of θ at point λ. In fact, the local Whittle likelihood estimation was originally developed by Künsch (1987) as an estimating approach for spectra around the origin, with the assumption that the process is long-memory and its spectral density behaves like $G\lambda^{-2d}$ as $\lambda \to 0+$. For the case of fractional process when $d \in [-1, 1]$, the asymptotic properties of the local Whittle estimator were developed by Shimotsu and Phillips (2006). These can be treated as the cases of a local Whittle likelihood estimator if $\rho_\theta(\lambda)$ are considered as parametric spectral densities. Considering that the form of spectral density for many models of stochastic process is like the form of $|\eta_\theta(\lambda)|^p$ which is the pth order norm of a function $\eta_\theta(\lambda) \in L^p([-\pi, \pi])$, then it becomes an L^p-local Whittle likelihood estimator. Furthermore, if we consider it as Koenker's quantile score, we can introduce a local Whittle quantile estimator. Hence, the configuration is very general. This chapter highlights the effective asymptotic properties of our general local Whittle estimator, and presents a comparison with other estimators.

In Sect. 3.2, the assumptions for the process and kernel function are provided and the related lemma is introduced. In Sect. 3.3, the asymptotic variance of $\rho_{\hat{\theta}}(\lambda)$ is shown to be $O((Nh)^{-1})$ when $h \to 0$ and $Nh \to \infty$ as $N \to \infty$, where N denotes the sample size and h denotes the bandwidth. In Sect. 3.4, for the following special case $\rho_\theta(\lambda) = |1 - \phi_\theta(\lambda)|^{-p}/(2\pi)$, if $p = 2$, for $AR(q)$ model with that the variance of innovation process is 1, it is shown that $\int_{-\pi}^{\pi} |1 - \phi_{\hat{\theta}(\lambda)}(\lambda)|^2 f(\lambda)\,\mathrm{d}\lambda$ describes the mean square prediction error, which can be easily understood if we think of the form of AR model's spectral density, i.e., $\rho_\theta(\lambda) = |1 - \phi_\theta(\lambda)|^{-p}/(2\pi)$. If we use the usual exponentially weighted linear prediction, the corresponding response function is $\phi_\theta^E(\lambda) = \theta \mathrm{e}^{\mathrm{i}\lambda}$, i.e., AR(1) response. Then, the mean square prediction error is $E_1 = \int_{-\pi}^{\pi} |1 - \phi_{\hat{\theta}}^E(\lambda)|^2 f(\lambda)\,\mathrm{d}\lambda$. In this case, if we apply our local Whittle approach for AR(1)-response, then the response function is given by $\phi_\theta^{LW}(\lambda) = \theta(\lambda)\mathrm{e}^{\mathrm{i}\lambda}$, i.e., "frequency dependent" AR(1) coefficient. Then, from $\phi_\theta^{LW}(\lambda)$, we can construct the one-step predictor by $\hat{z}(N+1) = \sum_{j=1}^{N+1} \phi_{\hat{\theta}_h(\lambda)}^{LW}(\lambda_j^{N+1}) \exp\{-\mathrm{i}(N+1)(\lambda_j^{N+1})\} \times \mathrm{d}Z_N(\lambda_j)$, where $\mathrm{d}Z_N(\lambda) = (1/N)\sum_{k=1}^{N} z(k)\exp\{\mathrm{i}k\lambda\}$ and $\lambda_j^{N+1} = -\pi + 2\pi j/(N+1)$. It is shown that $|1 - \phi_{\hat{\theta}_h(\lambda)}^{LW}(\lambda)|^{-2} \to^p f(\lambda)$, and $|1 - \phi_{\hat{\theta}}^E(\lambda)|^{-2} \not\to^p f(\lambda)$, and thus we obtain that the mean square error of $\hat{z}(N+1)$ is smaller than E_1. Furthermore, we introduce the general class of $\rho_\theta(\lambda)$ which contains Koenker's quantile score function, i.e., $\rho_\tau(\lambda)^{-1} = \lambda(\tau - \mathbb{I}\{\lambda < 0\})$, where $\mathbb{I}\{\cdot\}$ is the indicator function. Thus, the τ-quantile frequency regression can be discussed if we use $\rho_\tau(\lambda - \theta' y)$ where y is an exogenous variable. In Sect. 3.5, we show that although the local Whittle estimator for the spectral density has the same consistent order as the smoothed periodogram estimator, for fixed h, the former has a potentially smaller bias which is coincident with the result in Hjort and Jones (1996). Furthermore, we show that the prediction error of local Whittle predictor becomes smaller than that of the usual exponentially weighted linear prediction when the sample size is sufficiently large.

3.2 Preliminaries

Let $\{z(n) : n \in \mathbb{Z}\}$ be a real-valued zero mean stochastic process that satisfies the following assumptions:

(a) $\{z(n)\}$ is 4th-order stationary which all moments are finite;
(b) the joint kth-order cumulant $Q_z(j_1, \ldots, j_{k-1})$ of $z(n), z(n + j_1), \ldots, z(n + j_{k-1})$ satisfies

$$\sum_{j_1, \ldots, j_{k-1}=-\infty}^{\infty} (1 + |j_1|)|Q_z(j_1, \ldots, j_{k-1})| < \infty$$

for $k = 2, 3, 4$.

Then the spectral density of $\{z(n), n \in \mathbb{Z}\}$ is given by

$$f(\omega) = (1/2\pi) \sum_{j=-\infty}^{\infty} Q_z(j) \exp\{\mathrm{i} j\omega\},$$

and in the following discussion, we assume that $f(\omega)$ is positive on $[-\pi, \pi]$. Let $\{z(1), z(2), \ldots, z(N)\}$ be an observed sample of $\{z(n)\}$, and denote the periodogram of $\{z(n)\}$ by

$$I_N(\omega) = (1/2\pi N)\left|\sum_{n=1}^{N} z(n) \exp\{\mathrm{i}n\omega\}\right|^2.$$

Let $K(\cdot)$ be a kernel function that satisfies:

Assumption 3.1 (i) $K(t)$ is a non-negative even function of bounded variation; (ii) $\int_{-\infty}^{\infty} K(t)\,\mathrm{d}t = 1$, $\int_{-\infty}^{\infty} t^2 K(t)\,\mathrm{d}t = \sigma_K^2 < \infty$, $\int_{-\infty}^{\infty} t^2 K^2(t)\,\mathrm{d}t < \infty$.

In fact, the usual kernel functions satisfy this assumption, for example, the uniform kernel function

$$K(t) = \begin{cases} 1/2, & |t| \le 1; \\ 0, & |t| > 1, \end{cases}$$

and the parabolic kernel function

$$K_1(t) = \begin{cases} (3/4)(1-t^2), & |t| \le 1; \\ 0, & |t| > 1. \end{cases}$$

A counter-example which dose not satisfy item (ii) of Assumption 2.1 can be obtained if we consider the density function of Cauchy distribution as the kernel function K. Clearly, $\int_{-\infty}^{\infty} t^2 K(t)\,\mathrm{d}t = \infty$. The following lemma is a result of Theorem 5.6.3 (Brillinger 2001).

Define $f^N(\lambda) = (2\pi/N)\sum_{j=1}^{N} K_h(\lambda - \lambda_j) I_N(\lambda_j)$, where $K_h(\cdot) = K(\cdot/h)/h$, $h > 0$ and $\lambda_j = -\pi + 2\pi j/N$ $(j = 1, \ldots, N)$. We have

Lemma 3.1 *Let $K(\lambda)$ satisfy Assumption 3.1 and $\{z(t), t \in \mathbb{Z}\}$ be a stochastic process that satisfies conditions (a) and (b). If $hN \to \infty$ and $h \to 0$ as $N \to \infty$, then*

$$\sqrt{Nh}(f^N(\lambda) - f(\lambda)) \to^d N(0, \Psi(\lambda)),$$

where

$$\Psi(\lambda) = \begin{cases} 2\pi \int_{-\infty}^{\infty} K^2(s)\,\mathrm{d}s f^2(\lambda), & \lambda \ne 0; \\ 4\pi \int_{-\infty}^{\infty} K^2(s)\,\mathrm{d}s f^2(\lambda), & \lambda = 0. \end{cases}$$

We introduce the local Whittle likelihood function $D_{h,\lambda}(\cdot, \cdot)$,

$$D_{h,\lambda}(\rho_\theta, f) = \int_{-\pi}^{\pi} K_h(\lambda - \omega)\{\log \rho_\theta(\omega) + \rho_\theta(\omega)^{-1} f(\omega)\}\,\mathrm{d}\omega,$$

where $\rho_\theta(\omega)$ is a score function with an unknown parameter $\theta \in \Theta \subset \mathbb{R}^T$. We assume that $\rho_\theta(\omega)$ satisfies the following assumption.

Assumption 3.2 For arbitrary given λ, we can find a set Θ such that

(i) Θ is a compact of $\mathbb{R}^T$, where T is fixed;
(ii) $\rho_\theta(\omega)$ can be continuously three times differentiated with respect to $\theta \in \Theta$ and absolutely continuous with respect to $\omega \in [-\pi, \pi]$, and $\rho_\theta(\lambda) > 0$ on $\Theta \times [-\pi, \pi]$;
(iii) $\log \rho_\theta(\lambda) + \rho_\theta(\lambda)^{-1} f(\lambda)$ attains its minimum with respect to θ at a unique point on Θ, denoted by $\theta_0(\lambda)$. □

Let us set $\rho_\theta(\lambda) = (1/2\pi)|1 - \theta e^{i\lambda}|^{-2}$(AR(1) spectral density). If the true spectral density is given by AR(1) with a, i.e., $f(\lambda) = (1/2\pi)|1 - a e^{i\lambda}|^{-2}$, then for arbitrary λ, $\theta_0(\lambda) = a$ which implies that $\theta_0(\lambda)$ does not vary with λ. But if the true spectral density $f(\lambda)$ is not AR(1), then $\theta_0(\lambda)$ is given by θ satisfying $|1-\theta e^{i\lambda}|^{-2}=2\pi f(\lambda)$. Evidently, $\theta_0(\lambda)$ depends on λ. This example suggests us to understand the form of $\theta_0(\lambda)$, which implies θ_0 may vary with λ, and from the procedure to obtain θ_0, we can see that $\rho_{\theta_0(\lambda)}(\lambda)\equiv f(\lambda)$.

We define $\theta_h(\lambda)\in\Theta$ as the minimizer of $D_{h,\lambda}(\rho_\theta, f)$ on the compact set Θ. Particularly, if the minimizer is not unique, we write $\theta_h(\lambda)\in\Theta$ as an arbitrary minimizer of $D_{h,\lambda}(\rho_\theta, f)$, i.e.,

$$\theta_h(\lambda)\in \arg\min_{\theta\in\Theta} D_{h,\lambda}(\rho_\theta, f).$$

Notice that $\rho_{\theta_h(\lambda)}(\lambda)$ can estimate the spectral density $f(\cdot)$ at point λ. To give an extreme example, set $\rho_\theta(\cdot)=\theta$ $(\theta>0)$. Then $(\frac{\partial}{\partial\theta})D_{h,\lambda}(\rho_\theta, f)=0$ yields $\theta_h(\lambda)=\int_{-\pi}^{\pi}K_h(\lambda-\omega)f(\omega)\,d\omega$. Evidently $\theta_h(\lambda)$ $(=\rho_{\theta_h(\lambda)}(\cdot)$, in this case$)\to f(\lambda)$ if $h\to 0$.

In practice we do not know $f(\lambda)$. Replacing $f(\lambda)$ by $I_N(\lambda)$, we introduce $\hat\theta_h(\lambda)$ as an estimator of $\theta_h(\lambda)$ which is computable, and defined by

$$\hat\theta_h(\lambda)\in \arg\min_{\theta\in\Theta} \hat D_{h,\lambda}(\rho_\theta, I_N),$$

where $\hat D_{h,\lambda}(\rho_\theta, I_N)=(2\pi/N)\sum_{j=1}^{N}\{\log\rho_\theta(\lambda_j)+\rho_\theta(\lambda_j)^{-1}I_N(\lambda_j)\}K_h(\lambda-\lambda_j)$. We assume the following:

Assumption 3.3

$$(N^{1/2}h)^{-1}+N^{1/5}h\to 0, \textit{as } N\to\infty.$$

Under Assumption 3.3, we see that $hN\to\infty$ and $h\to 0$ as $N\to\infty$.

3.3 Estimation Theory

In the previous section, we introduced $\hat\theta_h(\lambda)$ as an estimator of $\theta_h(\lambda)$. In this section, we investigate the consistency of $\hat\theta_h(\lambda)$. For this, the following lemma and theorems are needed.

Lemma 3.2 *Under Assumption 3.1, if $g(\omega)$ is a function of bounded variation on $[-\pi,\pi]$, then*

$$(2\pi/N)\sum_{j=1}^{N} g(\lambda_j)K_h(\lambda-\lambda_j)=\int_{-\pi}^{\pi} g(\omega)K_h(\lambda-\omega)\,d\omega+O(h^{-1}N^{-1}).$$

Proof. Given that $g(\omega)$ and $K_h(\omega)$ are of bounded variation, $g(\omega)K_h(\lambda-\omega)$ is also of bounded variation. If $G(\omega)=g(\omega)K_h(\lambda-\omega)$, then there exist two monotonic functions $G_1(\omega)$ and $G_2(\omega)$ such that $G(\omega) = G_1(\omega) - G_2(\omega)$. From the second mean value theorem for definite integrals, given G_i where $i = 1$ or $i = 2$ is monotonic function, there exists $x_i \in [\lambda_j, \lambda_{j+1}]$ such that

$$\int_{\lambda_j}^{\lambda_{j+1}} G_i(\omega)\,\mathrm{d}\omega = G_i(\lambda_j)\int_{\lambda_j}^{x_i}\mathrm{d}\omega + G_i(\lambda_{j+1})\int_{x_i}^{\lambda_{j+1}}\mathrm{d}\omega.$$

It is equivalent to that there exists $\varepsilon_i \in [0, 2\pi/N]$, such that

$$\begin{aligned}\int_{\lambda_j}^{\lambda_{j+1}} G_i(\omega)\,\mathrm{d}\omega &= G_i(\lambda_j)\varepsilon_i + G_i(\lambda_{j+1}) \times (2\pi/N - \varepsilon_i)\\ &= (2\pi/N)G_i(\lambda_{j+1} + \varepsilon_i(G_i(\lambda_j) - G_i(\lambda_{j+1}))\end{aligned}$$

where $i = 1$ or 2. Noticing that $G(\omega) = G_1(\omega) - G_2(\omega)$, we have

$$\begin{aligned}\int_{\lambda_j}^{\lambda_{j+1}} G(\omega)\,\mathrm{d}\omega = (2\pi/N)(G_1(\lambda_{j+1}) - G_2(\lambda_{j+1})) - \varepsilon_1(G_1(\lambda_{j+1})\\ - G_1(\lambda_j)) + \varepsilon_2(G_2(\lambda_{j+1}) - G_2(\lambda_j)).\end{aligned}$$

Given $\varepsilon_i \in [0, 2\pi/N]$, the following inequality holds:

$$\begin{aligned}(2\pi/N)G(\lambda_{j+1}) - (2\pi/N)(G_1(\lambda_{j+1}) - G_1(\lambda_j)) \le \int_{\lambda_j}^{\lambda_{j+1}} G(\omega)\,\mathrm{d}\omega\\ \le (2\pi/N)G(\lambda_{j+1}) + (2\pi/N)(G_2(\lambda_{j+1}) - G_2(\lambda_j)).\end{aligned}$$

Thus

$$\begin{aligned}\sum_{j=0}^{N-1}(2\pi/N)G(\lambda_{j+1}) - (2\pi/N)\sum_{j=0}^{N-1}(G_1(\lambda_{j+1}) - G_1(\lambda_j)) \le \int_{-\pi}^{\pi} G(\omega)\,\mathrm{d}\omega\\ \le \sum_{j=0}^{N-1}(2\pi/N)G(\lambda_{j+1}) + \sum_{j=0}^{N-1}(2\pi/N)(G_2(\lambda_{j+1}) - G_2(\lambda_j)).\end{aligned}$$

That is

$$\begin{aligned}\int_{-\pi}^{\pi} G(\omega)\,\mathrm{d}\omega - (2\pi/N)\sum_{j=0}^{N-1}(G_2(\lambda_{j+1}) - G_2(\lambda_j)) \le \sum_{j=0}^{N-1}(2\pi/N)G(\lambda_{j+1})\\ \le \int_{-\pi}^{\pi} G(\omega)\,\mathrm{d}\omega + \sum_{j=0}^{N-1}(2\pi/N)(G_1(\lambda_{j+1}) - G_1(\lambda_j)).\end{aligned}$$

Noticing that $G_1(\omega)$ and $G_2(\omega)$ are monotonic, the following inequality can be derived:

$$\int_{-\pi}^{\pi} G(\omega)\,\mathrm{d}\omega - (2\pi/N)(G_2(\pi) - G_2(-\pi)) \le \sum_{j=0}^{N-1}(2\pi/N)G(\lambda_{j+1})$$
$$\le \int_{-\pi}^{\pi} G(\omega)\,\mathrm{d}\omega + (2\pi/N)(G_1(\pi) - G_1(-\pi)).$$

Furthermore, set $G_1(\omega) \equiv (1/2)V_{-\pi}^{\omega}(G) + (1/2)G(\omega)$ and $G_2(\omega) \equiv (1/2)V_{-\pi}^{\omega}(G) - (1/2)G(\omega)$, where $V_a^b(G) \equiv \sup_{P\in\mathfrak{P}} \sum_{i=0}^{n_P-1} |G(\omega_{i+1}) - G(\omega_i)|$ and the supremum is taken over the set $\mathfrak{P} = \{P = \{\omega_0, \ldots, \omega_{n_P}\} | P$ is a partition of $[a, b]$, satisfying $\omega_i \le \omega_{i+1}$ for $0 \le i \le n_{P-1}\}$. Thus $G_1(\pi) - G_1(-\pi) = (1/2)V_{-\pi}^{\pi}(G) + (1/2)(G(\pi) - G(-\pi)) \le V_{-\pi}^{\pi}(G)$ and $G_2(\pi) - G_2(-\pi) = (1/2)V_{-\pi}^{\pi}(G) + (1/2)(G(-\pi) - G(\pi)) \le V_{-\pi}^{\pi}(G)$ which implies that

$$\int_{-\pi}^{\pi} g(\omega)K_h(\lambda - \omega)\,\mathrm{d}\omega - (2\pi/N)V_{-\pi}^{\pi}(G) \le (2\pi/N)\sum_{j=1}^{N} g(\lambda_j)K_h(\lambda - \lambda_j)$$
$$\le \int_{-\pi}^{\pi} g(\omega)K_h(\lambda - \omega)\,\mathrm{d}\omega + (2\pi/N)V_{-\pi}^{\pi}(G),$$

Observing that $V_{-\pi}^{\pi}(G) = h^{-1}V_{-\pi}^{\pi}(g(\omega)K(\frac{\lambda-\omega}{h}))$ and $K(\frac{\lambda-\omega}{h})$ is positive and bounded, we have that there exists $M > 0$ such that

$$V_{-\pi}^{\pi}(G) \le h^{-1}MV_{-\pi}^{\pi}(g).$$

Thus, the lemma holds. □

Theorem 3.1 *Let $\{z(t), t \in \mathbb{Z}\}$ be a stochastic process that satisfies conditions (a) and (b). Under Assumptions 3.1, 3.2, and 3.3, for any fixed λ,*

$$|\theta_h(\lambda) - \theta_0(\lambda)| \to 0;$$

and

$$|\hat{\theta}_h(\lambda) - \theta_0(\lambda)| \to^{P} 0,$$

as $N \to \infty$.

Proof. The second statement was proved by following the proof of Theorem 2.1 (Newey 1991). Let $\hat{D}_{h,\lambda}(\rho_\theta, I_N) = C_N + D_N$, $\lambda_j = -\pi + 2\pi j/N (j = 1, \ldots, N)$, where

$$C_N = (2\pi/N)\sum_{j=1}^{N} \log \rho_\theta(\lambda_j)K_h(\lambda - \lambda_j)$$

and

$$D_N = (2\pi/N) \sum_{j=1}^{N} \rho_\theta(\lambda_j)^{-1} I_N(\lambda_j) K_h(\lambda - \lambda_j).$$

Given that $\rho_\theta(\lambda)^{-1}$ is three times differentiable, and K is of bounded variation, we can see that, on $[-\pi, \pi]$, $\rho_\theta(\omega)^{-1} K_h(\lambda - \omega)$ is of bounded variation. Hence, from Lemma 3.2,

$$C_N = \int_{-\pi}^{\pi} \log \rho_\theta(\omega) K_h(\lambda - \omega)\, d\omega + O((Nh)^{-1}),$$

which implies that $C_N \to \log \rho_\theta(\lambda)$ as $N \to \infty$. From the results in Sect. 7.4 of Brillinger (2001), pp. 248–251, we have $D_N \to^p \rho_\theta(\lambda)^{-1} f(\lambda)$. Thus, we obtain

$$\hat{D}_{h,\lambda}(\phi_\theta, I_N) \to^p \log \rho_\theta(\lambda) + \rho_\theta(\lambda)^{-1} f(\lambda). \tag{3.1}$$

Similarly, we can also see that

$$C_N + E_N \to \log \rho_\theta(\lambda) + \rho_\theta(\lambda)^{-1} f(\lambda), \tag{3.2}$$

where $E_N = (2\pi/N) \sum_{j=1}^{N} \rho_\theta(\lambda_j)^{-1} f(\lambda_j) K_h(\lambda - \lambda_j)$.

Below we show that if $\lim_{N\to\infty} \sup_{\theta\in\Theta} |\hat{D}_{h,\lambda}(\rho_\theta, I_N) - \log \rho_\theta(\lambda) + \rho_\theta(\lambda)^{-1} f(\lambda)| = 0$, then the second item of theorem holds and similarly, the first item can be proved. It can be observed that on the compact set $\Theta \times [-\pi, \pi]$, $\rho_\theta(\lambda)$ is continuous with respect to θ and λ. Hence, for an arbitrary $\varepsilon > 0$, there exists $\delta > 0$, such that, for an arbitrary θ_1 and θ_2 satisfying $|\theta_1 - \theta_2| < 2\delta$, $|\rho_{\theta_1}(\lambda_j)^{-1} - \rho_{\theta_2}(\lambda_j)^{-1}| < \varepsilon/(4f(\lambda))$ holds. Additionally, we define (θ, δ) as the open set $\{\theta' : |\theta' - \theta| < \delta\}$. Then for an open covering $\cup_\theta \ (\theta, \delta) \supset \Theta$, there exists an open subcovering $(\theta_l, \delta)_{l=1}^{L}$. Thus,

$$\begin{aligned}
&\sup_{\theta\in\Theta} |C_N + D_N - C_N - E_N| \\
&= \sup_{\theta\in\Theta} (2\pi/N) \Big| \sum_{j=1}^{N} K_h(\lambda - \lambda_j) \rho_\theta(\lambda_j)^{-1} (I_N(\lambda_j) - f(\lambda_j)) \Big| \\
&\le \max_{l} (2\pi/N) \Big| \sum_{j=1}^{N} K_h(\lambda - \lambda_j) \rho_{\theta_l}(\lambda_j)^{-1} (I_N(\lambda_j) - f(\lambda_j)) \Big| + \\
&\sup_{l, \theta \in \ (\theta_l, \delta)} (2\pi/N) \sum_{j=1}^{N} K_h(\lambda - \lambda_j) \Big| \rho_{\theta_l}(\lambda_j)^{-1} - \rho_\theta(\lambda_j)^{-1} \Big| \times (I_N(\lambda_j) + f(\lambda_j)).
\end{aligned}$$

Then from Lemma 3.1,

$$(2\pi/N) \Big| \sum_{j=1}^{N} K_h(\lambda - \lambda_j) \rho_{\theta_l}(\lambda_j)^{-1} (I_N(\lambda_j) - f(\lambda_j)) \Big| \to^p 0,$$

and

$$(2\pi/N)\sum_{j=1}^{N} K_h(\lambda-\lambda_j)\Big|\rho_{\theta_l}(\lambda_j)^{-1}-\rho_\theta(\lambda_j)^{-1}\Big|$$
$$\times (I_N(\lambda_j)+f(\lambda_j)) \to^p (\varepsilon/4f(\lambda))2f(\lambda),$$

as $N \to \infty$. Thus

$$Pr(\sup_{\theta\in\Theta}|C_N+D_N-C_N-E_N|>2\varepsilon)$$
$$\le Pr(o_p(1)>\varepsilon)+Pr(\varepsilon/(4f(\lambda))(2f(\lambda)+o_p(1))>\varepsilon)\to 0,$$

as $N \to \infty$, which implies that $\lim_{N\to\infty}\sup_{\theta\in\Theta}|C_N+D_N-C_N-E_N| = 0$ in probability. Furthermore, from (3.2), in terms of probability, the following can be derived

$$\lim_{N\to\infty}\sup_{\theta\in\Theta}\Big|\hat{D}_{h,\lambda}(\rho_\theta, I_N)-\log\rho_\theta(\lambda)-\rho_\theta(\lambda)^{-1}f(\lambda)\Big| = 0. \tag{3.3}$$

This equation implies that $|\min_\theta \hat{D}_{h,\lambda}(\rho_\theta, I_N)-\min_\theta\{\log\rho_\theta(\lambda)+\rho_\theta(\lambda)^{-1}f(\lambda)\}|$ $\to^p 0$, i.e.,

$|\hat{D}_{h,\lambda}(\rho_{\hat{\theta}_h(\lambda)}, I_N)-\log\rho_{\theta_0}(\lambda)-\rho_{\theta_0}(\lambda)^{-1}f(\lambda)| \to^p 0$. Here we use proof by contradiction to prove it. If it does not hold, it means that the $|\min_\theta \hat{D}_{h,\lambda}(\rho_\theta, I_N) - \min_\theta\{\log\rho_\theta(\lambda)+\rho_\theta^{-1}(\lambda)f(\lambda)\}| \to^p 0$ does not hold, which means that there exist $\varepsilon > 0$ and $\delta > 0$, such that for arbitrary n, there exists $N > n$, such that $Pr(|\min_\theta \hat{D}_{h,\lambda}(\rho_\theta, I_N)-\min_\theta\{\log\rho_\theta(\lambda)+\rho_\theta^{-1}(\lambda)f(\lambda)\}|>\varepsilon)>\delta$. In the situation when $|\min_\theta \hat{D}_{h,\lambda}(\rho_\theta, I_N)-\min_\theta\{\log\rho_\theta(\lambda)+\rho_\theta^{-1}(\lambda)f(\lambda)\}|>\varepsilon$, if $\min_\theta \hat{D}_{h,\lambda}$ $> \min_\theta\{\log\rho_\theta(\lambda)+\rho_\theta^{-1}(\lambda)f(\lambda)\}$, it implies that $\hat{D}_{h,\lambda}(\rho_{\theta_0}, I_N)-\{\log\rho_{\theta_0}(\lambda)+\rho_{\theta_0}^{-1}$ $f(\lambda)\}>\varepsilon$, if not, it implies that $\{\log\rho_{\hat{\theta}_h}+\rho_{\hat{\theta}_h}^{-1}f(\lambda)\}-\hat{D}_{h,\lambda}(\rho_{\hat{\theta}_h}, I_N)>\varepsilon$. Thus $Pr(\Big|\hat{D}_{h,\lambda}(\rho_{\hat{\theta}_h(\lambda)}, I_N)-\log\rho_{\hat{\theta}_h(\lambda)}(\lambda)-\rho_{\hat{\theta}_h(\lambda)}(\lambda)^{-1}f(\lambda)\Big|>\varepsilon)+Pr(\Big|\hat{D}_{h,\lambda}(\rho_{\theta_0}, I_N)$ $-\log\rho_{\theta_0}(\lambda)-\rho_{\theta_0}(\lambda)^{-1}f(\lambda)\Big|>\varepsilon)\ge Pr(|\min_\theta \hat{D}_{h,\lambda}(\rho_\theta, I_N)-\min_\theta\{\log\rho_\theta(\lambda)+$ $\rho_\theta(\lambda)^{-1}f(\lambda)\}|>\varepsilon)>\delta$ holds. However, from expression (3), $Pr(\Big|\hat{D}_{h,\lambda}(\rho_{\hat{\theta}_h(\lambda)}, I_N)$ $-\log\rho_{\hat{\theta}_h(\lambda)}(\lambda)-\rho_{\hat{\theta}_h(\lambda)}(\lambda)^{-1}f(\lambda)\Big|>\varepsilon)\to 0$ and $Pr(\Big|\hat{D}_{h,\lambda}(\rho_{\theta_0}, I_N)-\log\rho_{\theta_0}$ $(\lambda)-\rho_{\theta_0}(\lambda)^{-1}f(\lambda)\Big|>\varepsilon)\to 0$. That is $Pr(\Big|\hat{D}_{h,\lambda}(\rho_{\hat{\theta}_h(\lambda)}, I_N)-\log\rho_{\hat{\theta}_h(\lambda)}(\lambda)-$ $\rho_{\hat{\theta}_h(\lambda)}(\lambda)^{-1}f(\lambda)\Big|>\varepsilon)+Pr(\Big|\hat{D}_{h,\lambda}(\rho_{\theta_0}, I_N)-\log\rho_{\theta_0}(\lambda)-\rho_{\theta_0}(\lambda)^{-1}f(\lambda)\Big|>\varepsilon)\to$ 0. It leads to a contradiction. Furthermore, we have the following: $|\log\rho_{\hat{\theta}_h}(\lambda)+$ $\rho_{\hat{\theta}_h}(\lambda)^{-1}f(\lambda)-\log\rho_{\theta_0}(\lambda)-\rho_{\theta_0}(\lambda)^{-1}f(\lambda)| \to^p 0$ by noticing that $|\hat{D}_{h,\lambda}(\rho_{\hat{\theta}_h(\lambda)}, I_N)$ $-\log\rho_{\hat{\theta}_h}(\lambda)-\rho_{\hat{\theta}_h}(\lambda)^{-1}f(\lambda)| \to^p 0$. Given that $\arg\min_\theta\{\log\rho_\theta(\lambda)+\rho_\theta(\lambda)^{-1}f(\lambda)\}$ $=\theta_0(\lambda)$ is unique, we have $\hat{\theta}_h(\lambda)\to^p \theta_0(\lambda)$. Similarly, we can show that $\theta_h(\lambda)\to$ $\theta_0(\lambda)$ as $h \to 0$. □

From Theorem 3.1, we obtain the consistency of $\hat{\theta}_h(\lambda)$ and $\theta_h(\lambda)$. Observing that $\rho_{\theta_0(\lambda)}(\lambda) = f(\lambda)$ enables us to investigate the asymptotic behavior of $\rho_{\hat{\theta}_h}(\lambda)$.

Theorem 3.2 *Let $\{z(t), t \in \mathbb{Z}\}$ be a stochastic process that satisfies conditions (a) and (b). Under Assumptions 3.1, 3.2 and 3.3,*

$$\sqrt{Nh}(\rho_{\hat{\theta}_h}(\lambda) - f(\lambda)) \to^d N(0, \Psi(\lambda)),$$

as $N \to \infty$ where Ψ is defined as Lemma 3.1.

Proof. From the definition of $\hat{\theta}_h(\lambda)$ and $\theta_h(\lambda)$, we have the following:

$$\frac{\partial \hat{D}_{h,\lambda}(\rho_\theta, I_N)}{\partial \theta}|_{\theta=\hat{\theta}_h(\lambda)} = (2\pi/N)\sum_{j=1}^{N} K_h(\lambda - \lambda_j)$$
$$\{I_N(\lambda_j)\frac{\partial \rho_\theta(\lambda_j)^{-1}}{\partial \theta} + \frac{1}{\rho_\theta(\lambda_j)}\frac{\partial \rho_\theta(\lambda_j)}{\partial \theta}\}|_{\theta=\hat{\theta}_h(\lambda)} = 0, \tag{3.4}$$

$$\frac{\partial D_{h,\lambda}(\rho_\theta, f)}{\partial \theta}|_{\theta=\theta_h(\lambda)} = \int_{-\pi}^{\pi} K_h(\lambda - \omega)$$
$$\{f(\omega)\frac{\partial \rho_\theta(\omega)^{-1}}{\partial \theta} + \frac{1}{\rho_\theta(\omega)}\frac{\partial \rho_\theta(\omega)}{\partial \theta}\}|_{\theta=\theta_h(\lambda)}\, d\omega = 0. \tag{3.5}$$

By Taylor series expansion, (3.4) can be written as

$$\begin{aligned}\frac{\partial \hat{D}_{h,\lambda}(\rho_\theta, I_N)}{\partial \theta}|_{\theta=\hat{\theta}_h(\lambda)} &= (2\pi/N)\sum_{j=1}^{N} K_h(\lambda - \lambda_j)\\ &\quad\{I_N(\lambda_j)\frac{\partial \rho_\theta(\lambda_j)^{-1}}{\partial \theta} + \frac{1}{\rho_\theta(\lambda_j)}\frac{\partial \rho_\theta(\lambda_j)}{\partial \theta}\}|_{\theta=\theta_h(\lambda)}\\ &\quad + \frac{\partial^2}{\partial\theta\partial\theta'}\hat{D}_{h,\lambda}(\rho_\theta, I_N)|_{\theta=\theta_N}(\hat{\theta}_h(\lambda) - \theta_h(\lambda)) = 0,\end{aligned} \tag{3.6}$$

where θ_N is between $\theta_h(\lambda)$ and $\hat{\theta}_h(\lambda)$. Then, from Lemma 3.2, the left-hand side of (3.5) can be written as follows:

$$\begin{aligned}\frac{\partial D_{h,\lambda}(\rho_\theta, f)}{\partial \theta}|_{\theta=\theta_h(\lambda)} &= (2\pi/N)\sum_{j=1}^{N} K_h(\lambda - \lambda_j)\\ &\quad\{f(\lambda_j)\frac{\partial \rho_\theta(\lambda_j)^{-1}}{\partial \theta} + \frac{1}{\rho_\theta(\lambda_j)}\frac{\partial \rho_\theta(\lambda_j)}{\partial \theta}\}|_{\theta=\theta_h(\lambda)}\\ &\quad + O(h^{-1}N^{-1}) = 0.\end{aligned} \tag{3.7}$$

Combining (3.6) and (3.7), we have

$$(2\pi/N)\sum_{j=1}^{N} K_h(\lambda-\lambda_j)(I_N(\lambda_j)-f(\lambda_j))\frac{\partial \rho_\theta(\lambda_j)^{-1}}{\partial\theta}|_{\theta=\theta_h(\lambda)}+$$
$$\frac{\partial^2}{\partial\theta\partial\theta'}\hat{D}_{h,\lambda}(\rho_\theta, I_N)|_{\theta=\theta_N}(\hat{\theta}_h(\lambda)-\theta_h(\lambda)) = O(N^{-1}h^{-1}).$$

From Theorem 3.1, $|\hat{\theta}_h(\lambda)-\theta_h(\lambda)| \to^p 0$. Thus, $(\frac{\partial^2}{\partial\theta\partial\theta'}\hat{D}_{h,\lambda}(\rho_\theta, I_N))|_{\theta=\theta_N}$ $= (\frac{\partial^2}{\partial\theta\partial\theta'}\hat{D}_{h,\lambda}(\rho_\theta, I_N))|_{\theta=\theta_h} + \mathbf{o}_p(1)$. Then, we have

$$[\frac{\partial^2}{\partial\theta\partial\theta'}\hat{D}_{h,\lambda}(\rho_{\theta_h(\lambda)}, I_N)+\mathbf{o}_p(1)]\sqrt{Nh}(\hat{\theta}_h(\lambda)-\theta_h(\lambda)) =$$
$$\sqrt{Nh}(2\pi/N)\sum_{j=1}^{N} K_h(\lambda-\lambda_j)(I_N(\lambda_j)-f(\lambda_j))\frac{\partial \rho_\theta(\lambda_j)^{-1}}{\partial\theta}|_{\theta=\theta_h(\lambda)} + O(h^{-1}N^{-1})\sqrt{Nh}$$

$$\Leftrightarrow$$

$$[M_{N,h}(\lambda)+\mathbf{o}_p(1)]\sqrt{Nh}(\hat{\theta}_h(\lambda)-\theta_h(\lambda)) = \Delta_{N,h}(\lambda) + O(h^{-1/2}N^{-1/2}),$$

where

$$M_{N,h}(\lambda) = \frac{\partial^2}{\partial\theta\partial\theta'}\hat{D}_{h,\lambda}(\rho_{\theta_h(\lambda)}, I_N),$$

and

$$\Delta_{N,h}(\lambda) = \sqrt{Nh}(2\pi/N)\sum_{j=1}^{N} K_h(\lambda-\lambda_j)(I_N(\lambda_j)-f(\lambda_j))\frac{\partial \rho_\theta(\lambda_j)^{-1}}{\partial\theta}|_{\theta=\theta_0(\lambda)}.$$

As $h \to 0$, by following the line of the proof of Theorem 5.6.3 (Brillinger 2001) we have

$$M_{N,h}(\lambda)+\mathbf{o}_p(1) \to^p M(\lambda),$$

and

$$\Delta_{N,h}(\lambda) \to^d N(0, V_0)$$

where $M(\lambda) = f^2(\lambda)\frac{\partial\rho_\theta^{-1}}{\partial\theta}\frac{\partial\rho_\theta^{-1}}{\partial\theta'}|_{\theta=\theta_0(\lambda)}$, and

$$V_0 = \begin{cases} 2\pi\int_{-\infty}^{\infty} K^2(s)\,\mathrm{d}s f^2(\lambda)\frac{\partial\rho_\theta(\lambda)^{-1}}{\partial\theta}\frac{\partial\rho_\theta(\lambda)^{-1}}{\partial\theta'}|_{\theta=\theta_0(\lambda)}, & \lambda \neq 0; \\ 4\pi\int_{-\infty}^{\infty} K^2(s)\,\mathrm{d}s f^2(\lambda)\frac{\partial\rho_\theta(\lambda)^{-1}}{\partial\theta}\frac{\partial\rho_\theta(\lambda)^{-1}}{\partial\theta'}|_{\theta=\theta_0(\lambda)}, & \lambda = 0. \end{cases}$$

Let $M(\lambda)^-$ be the generalized inverse of $M(\lambda)$, i.e.,

$$M(\lambda)^- = \frac{f^{-2}(\lambda)}{||\frac{\partial\rho_\theta^{-1}}{\partial\theta}||^4}\frac{\partial\rho_\theta^{-1}(\lambda)}{\partial\theta}\frac{\partial\rho_\theta^{-1}(\lambda)}{\partial\theta'}|_{\theta=\theta_0(\lambda)}.$$

Thus, $\sqrt{Nh}(\hat{\theta}_h(\lambda) - \theta_h(\lambda))$ is asymptotically normal with $N(0, M(\lambda)^- V_0 M(\lambda)^-)$. Furthermore, for a given smooth function g, it can be seen that as $h \to 0$,

$$\int K_h(\lambda - \omega)g(\omega)\,\mathrm{d}\omega = g(\lambda) + 1/2\sigma_K^2 h^2 g''(\lambda) + O(h^4)$$

by using Taylor series expansion. Then, from (3.5) we have

$$\begin{aligned}
0 &= \frac{\partial D_{h,\lambda}(\rho_\theta, f)}{\partial\theta}|_{\theta=\theta_h(\lambda)}\\
&= \int_{-\pi}^{\pi} K_h(\lambda-\omega)\{f(\omega)\frac{\partial\rho_\theta(\omega)^{-1}}{\partial\theta} + \frac{1}{\rho_\theta(\omega)}\frac{\partial\rho_\theta(\omega)}{\partial\theta}\}|_{\theta=\theta_h(\lambda)}\,\mathrm{d}\omega\\
&= \rho_\theta(\lambda)^{-2}\frac{\partial}{\partial\theta}\rho_\theta(\lambda)(\rho_\theta(\lambda) - f(\lambda)) + 1/2h^2\sigma_K^2\frac{\partial^2}{\partial\lambda^2}\\
&\quad \{\rho_\theta(\lambda)^{-2}\frac{\partial}{\partial\theta}\rho_\theta(\lambda)(\rho_\theta(\lambda) - f(\lambda))\} + o(h^4)|_{\theta=\theta_h(\lambda)},
\end{aligned}$$

which implies that

$$\rho_{\theta_h}(\lambda) - f(\lambda) = o(h^2) = o(N^{-1/2}h^{-1/2}),$$

from Assumption 3.3. Thus,

$$\begin{aligned}
&\sqrt{Nh}(\rho_{\hat{\theta}_h}(\lambda) - f(\lambda)) = \sqrt{Nh}(\rho_{\hat{\theta}_h}(\lambda)^{-1} - \rho_{\theta_h}(\lambda) + \rho_{\theta_h} - f(\lambda))\\
&= \sqrt{Nh}(\rho_{\hat{\theta}_h}(\lambda) - \rho_{\theta_h}(\lambda)) + o(1).
\end{aligned}$$

Furthermore, as

$$\sqrt{Nh}(\rho_{\hat{\theta}_h}(\lambda) - \rho_{\theta_h}(\lambda)) = \sqrt{Nh}(\hat{\theta}_h(\lambda) - \theta_h(\lambda))\frac{\partial\rho_\theta}{\partial\theta}|_{\theta=\theta_N},$$

by the convergence of $\sqrt{Nh}(\hat{\theta}_h(\lambda) - \theta_h(\lambda))$, we can get the convergence of $\sqrt{Nh}$ $(\rho_{\hat{\theta}_h}(\lambda) - \rho_{\theta_h}(\lambda))$. Furthermore, we can obtain that:

$$\sqrt{Nh}(\rho_{\hat{\theta}_h}(\lambda) - f(\lambda)) \to^d N(0, \Psi(\lambda)),$$

as $N \to \infty$. □

From Theorem 3.2 and Lemma 3.1, we know that the asymptotic variance is $O((Nh)^{-1})$, which is the same order as that of smoothed periodogram estimator.

3.4 Applications of the Local Whittle Estimator

In this section, we introduce two applications of the local Whittle estimator. The first application is one-step ahead prediction. A comparison between the prediction using the local Whittle estimator and prediction using the usual exponentially weighted linear predictor (**Cox**) is presented below. Let $\phi_\theta(\lambda) = \theta \exp\{\mathrm{i}\lambda\}$ and $\{z(k)\}$ satisfy an AR(q) model in which the variance of disturbance is 1 and q is unknown. When $\rho_\theta(\lambda) = |1 - \phi_\theta(\lambda)|^{-2}/(2\pi)$, we consider the one-step predictor $\hat{z}(N+1) = \sum_{j=1}^{N+1} \phi_{\hat{\theta}_h(\lambda_j^{N+1})}(\lambda_j^{N+1}) \exp\{-\mathrm{i}(N+1)(\lambda_j^{N+1})\}\Delta Z_N(\lambda_j^{N+1})$, where $\Delta Z_N(\lambda) = (1/N)\sum_{k=1}^{N} z(k)\exp\{\mathrm{i}k\lambda\}$ and $\lambda_j^{N+1} = -\pi + 2\pi j/(N+1)$.

From Brillinger (2001), we have $\lim_{N\to\infty} cov\{\Delta Z_N(\lambda), \overline{\Delta Z_N(\mu)}\} = \eta(\lambda-\mu) f(\lambda)\,\mathrm{d}\mu$, where $\eta(\lambda)$ is the period 2π extension of the Dirac delta function. We then write $\mathrm{d}Z(\lambda) \equiv l.i.m._{N\to\infty}\Delta Z_N(\lambda)$. Then, $z(t) = \int_{-\pi}^{\pi} \exp\{-\mathrm{i}t\lambda\}\,\mathrm{d}Z(\lambda)$, and

$$l.i.m._{n\to\infty} \sum_{j=1}^{N+1} \phi_{\hat{\theta}_h(\lambda_j^{N+1})}(\lambda_j^{N+1})\Delta Z_N(\lambda_j^{N+1})) = \int_{-\pi}^{\pi} \exp\{\mathrm{i}\lambda\}\theta_0(\lambda)\,\mathrm{d}Z(\lambda).$$

Thus, the mean squared prediction error (MSPE)

$$\begin{aligned}
&\lim_{N\to\infty} MSPE|z(N+1) - \hat{z}(N+1)|^2 = E\left|\int_{-\pi}^{\pi} (1 - \exp\{\mathrm{i}\lambda\}\theta_0(\lambda))\,\mathrm{d}Z(\lambda)\right|^2 \\
&= \int_{-\pi}^{\pi} |1 - \exp\{\mathrm{i}\lambda\}\theta_0(\lambda)|^2 f(\lambda)\,\mathrm{d}\lambda \\
&= 1.
\end{aligned}$$

From Grenander and Rosenblatt (1957), we know that the MSPE of this predictor converges to the minimum prediction error of the usual linear combination of form $\tilde{z}(t) = \sum_{j\geq 1} b_j z(t-j)$. If we use the usual exponentially weighted linear prediction to estimate the spectral density, then $f(\lambda) = |\sum_{j=0}^{\infty} \theta^j \mathrm{e}^{\mathrm{i}j\lambda}|^2/(2\pi)$. Observe that $|\theta| < 1$, and that

$$|\sum_{j=0}^{\infty} \theta^j \mathrm{e}^{\mathrm{i}j\lambda}|^2/(2\pi) = |1 - \theta\mathrm{e}^{\mathrm{i}\lambda}|^{-2}/(2\pi).$$

This means the fitting model is AR(1); thus, the one-step ahead predictor can be expressed as

$$\tilde{z}(N+1) = \tilde{\theta} \times z(N).$$

Given that for the true model AR(q), when $q \neq 1$, the MSPE of $\tilde{z}(N+1)$ is greater than 1, the former predictor would become better as the sample size increases. The second application of local Whittle estimator is that the score function $\rho_\theta(\lambda)$ could be a quantile loss function. Let $\rho_\theta(\lambda) = (\lambda - \theta'\mathbf{y})(\tau - I(\lambda - \theta'\mathbf{y} < 0))$, where θ is p-dimensional parameter, $\mathbf{y}$ is a given p-dimensional exogenous variable, and τ is given.

Assume that $\lambda - \theta'\mathbf{y} > 0$, $\rho_\theta(\lambda) = (\lambda - \theta'\mathbf{y})\tau$ is three times differentiable with respect to θ, and that there are p given exogenous variables, $\mathbf{y}_1, \mathbf{y}_2 \ldots,$ and $\mathbf{y}_p$. Corresponding to these exogenous variables, we can obtain $\hat{\theta}_h$ which is the minimizer of all of the following expressions.

$$\hat{D}_{h,\lambda}(\rho_\theta, I_N) = (2\pi/N)\sum_{j=1}^{N}\{\log \rho_\theta(\lambda_j) + \rho_\theta^{-1}(\lambda_j)I_N(\lambda_j)\}K_h(\lambda - \lambda_j),$$

$$\rho_\theta(\lambda) = (\lambda - \theta'\mathbf{y}_i)(\tau - I(\lambda - \theta'\mathbf{y}_i < 0)),$$

where $i = 1, 2, \ldots, p$. Thus, we can discuss τ-quantile frequency regression. That is, $\rho_\theta(\cdot)$ detects τ-quantile regression of Fourier components $d(\lambda_j) \equiv (\sqrt{2\pi N})^{-1}\sum_{n=1}^{N} z(n)\exp(in\lambda;\, \tau)$ on the exogenous variables $\mathbf{y}_1, \mathbf{y}_2 \ldots,$ and $\mathbf{y}_p$.

3.5 Numerical Results

Consider the following MA(3) model:

$$z(t) = \varepsilon(t) + 0.1\varepsilon(t-1) + 0.3\varepsilon(t-2) - 0.4\varepsilon(t-3)$$

where $\varepsilon(t) \sim N(0, 1)$. We compared the bias of the local Whittle likelihood estimator and smoothed periodogram estimator with the following kernel functions. Let $\rho_\theta(\lambda)^{-1} = |1 - \phi_\theta(\lambda)|^{-2.5}$, $\phi_\theta(\lambda) = \theta e^{i\lambda}$, and

$$K_1(\lambda) = \begin{cases} 1 - |\lambda|, & |\lambda| \leq 1; \\ 0, & |\lambda| > 1, \end{cases}, \quad (Bartlett)$$

$$K_2(\lambda) = \begin{cases} (1/2)(1 + \cos(\pi\lambda)), & |\lambda| \leq 1; \\ 0, & |\lambda| > 1, \end{cases}, \quad (Tukey - Hanning)$$

and $K_3(\lambda) = \exp(-2|\lambda|)$, $(Abel)$. Then, using 500 simulations, $N = 1000$, and $h = 0.1$, we obtained the results listed in Table 3.1.

The results seen in Table 3.1 show that for kernel function $K_1(\lambda)$, local Whittle likelihood estimator has a smaller bias than smoothed periodogram estimator, except for $K_2(\lambda)$, at point $4/5\pi$, and $K_3(\lambda)$, at $2/5\pi$ and $4/5\pi$. We can establish the follow-

Table 3.1 Bias and variance of local Whittle likelihood approach and smoothed periodogram approach

		$\lvert 1-\phi_{\theta_{\hat{h}}}\rvert_{K_1}^{-p}$	$\hat{f}_{K_1}$	$\lvert 1-\phi_{\theta_{\hat{h}}}\rvert_{K_2}^{-p}$	$\hat{f}_{K_2}$	$\lvert 1-\phi_{\theta_{\hat{h}}}\rvert_{K_3}^{-p}$	$\hat{f}_{K_3}$
0	Bias	0.00260676	0.00270855	0.00258221	0.00269828	0.00044157	0.00069107
	Variance	0.00208813	0.00208352	0.00235253	0.00234733	0.00162781	0.00161932
$1/5\pi$	Bias	0.00003677	0.00011719	0.00007258	0.00019083	0.00125056	0.00137813
	Variance	0.00331743	0.00330992	0.00365373	0.00365101	0.00250493	0.00250708
$2/5\pi$	Bias	0.00012137	0.00058333	0.00045973	0.00075678	0.00180937	0.00035954
	Variance	0.00225756	0.00227556	0.00249236	0.00250877	0.00172447	0.00175470
$3/5\pi$	Bias	0.00111715	0.00116949	0.00122641	0.00125814	0.00330884	0.00347701
	Variance	0.00004105	0.00004125	0.00004600	0.00004613	0.00003269	0.00003306
$4/5\pi$	Bias	0.00101466	0.00113484	0.00390239	0.00377189	0.00348700	0.00328538
	Variance	0.00146575	0.00146156	0.00189411	0.00188784	0.00128034	0.00127424

Table 3.2 Prediction errors of local Whittle estimator and exponentially weighted linear predictor

Sample size	200	400	800	1000
Local Whittle	1.2736	0.7745	0.8886	1.0063
Exponentially weighted linear	1.0093	0.7876	0.9059	1.0332

ing: $\rho_\theta(\lambda)$ has a potentially smaller bias than $\hat{f}_K$ and the superiority of the former depends on the kernel functions when considering different kernel functions.

Next, we consider the AR(2) model,

$$z(t) = 0.5z(t-1) + 0.1z(t-2) + \varepsilon(t)$$

where $\varepsilon(t) \sim N(0, 1)$. As the sample size increases gradually (from 200 to 1000), we compare prediction using the local Whittle estimator

$$\hat{z}(N+1) = \sum_{j=1}^{N+1} \phi_{\hat{\theta}_h}(-\pi + \frac{2\pi j}{N+1}) \exp\{-i(N+1)(-\pi + \frac{2\pi j}{N+1})\}$$
$$\Delta Z_N(-\pi + \frac{2\pi j}{N+1}),$$

where $\Delta Z_N(\lambda) = (1/N)\sum_{k=1}^{N} z(k)\exp\{ik\lambda\}$ with the usual exponentially weighted linear prediction

$$\tilde{z}(N+1) = \tilde{\theta} \times z(N),$$

which is obtained as the best linear predictor when the fitting spectral density $f_\theta(\lambda) = \frac{1}{2\pi}|1 - \theta\exp\{i\lambda\}|^{-2}/(2\pi)$. Table 3.2 lists the numerical results.

From Table 3.2, it can be seen that for sample sizes greater or equal 400, the MSPE of local Whittle estimator becomes smaller than that of exponentially weighted linear predictor.

References

Brillinger, D. R. (2001). *Time Series: Data Analysis and Theory*. Expanded. San Francisco: Holden-Day.

Grenander, U. and M. Rosenblatt (1957). *Statistical Analysis of Stationary Time Series*. New York: John Wiley & Sons.

Hannan, E. J. (1970). *Multiple Time Series*. New York: Wiley.

Hjort, Nils Lid and M Chris Jones (1996). "Locally parametric nonparametric density estimation". In: *The Annals of Statistics*, pp. 1619–1647.

Künsch, H. (1987). "Statistical aspects of self-similar processed". In: *In: Prokhorov, Yu., Sazanov, V.V. (Eds.), Proceedings of the First World Congress of the Bernoulli Societ* 1, pp. 67–74.

Liu, Regina Y, Kesar Singh, et al. (1992). "Moving blocks jackknife and bootstrap capture weak dependence". In: *Exploring the limits of bootstrap* 225, p. 248.

Newey, Whitney K (1991). "Uniform convergence in probability and stochastic equicontinuity". In: *Econometrica: Journal of the Econometric Society*, pp. 1161–1167.

Politis, Dimitris N and Joseph P Romano (1992). "A general resampling scheme for triangular arrays of α-mixing random variables with application to the problem of spectral density estimation". In: *The Annals of Statistics*, pp. 1985–2007.

Shimotsu, Katsumi and Peter CB Phillips (2006). "Local Whittle estimation of fractional integration and some of its variants". In: *Journal of Econometrics* 130.2, pp. 209–233.

Taniguchi, Masanobu (1980). "On estimation of the integrals of certain functions of spectral density". In: *Journal of Applied Probability* 17.1, pp. 73–83.

Taniguchi, Masanobu (1987). "Minimum contrast estimation for spectral densities of stationary processes". In: *Journal of the Royal Statistical Society: Series B (Methodological)* 49.3, pp. 315–325.

Tjøstheim, Dag and Karl Ove Hufthammer (2013). "Local Gaussian correlation: A new measure of dependence". In: *Journal of Econometrics* 172.1, pp. 33–48.

Xue, Yujie and Masanobu Taniguchi (2020). "Local Whittle likelihood approach for generalized divergence". In: *Scandinavian Journal of Statistics* 47.1, pp. 182–195.

Chapter 4
Systemic Risk in Energy Markets: Measuring Co-movements in Energy Asset Prices During Crises

Abstract This chapter develops forward-looking measures to quantify the exposure of the non-energy sector to extreme energy price shocks. Estimation is based on a vector-error correction model and multiplicative GARCH framework with a dynamic principal component analysis, capturing cointegration, volatility spillovers, and tail dependence with a latent energy market factor. The methodology is applied to energy futures traded on the European Energy Exchange and to DAX industrial returns, highlighting the transmission of energy stress during financial crises and major energy market events.

Keywords Energy crisis · Factor models · Marginal expected shortfall · Market integration

4.1 Introduction

In this chapter, a methodology is developed to quantify the exposure of the economy to extreme energy price shocks.The objective is to estimate the losses incurred by the non-energy sector in the event of a severe energy price shock, and to identify which energy assets contribute most to those losses. The resulting measure captures the expected shortfall for non-energy firms conditional on a severe stress event in energy markets.

To construct this measure, the conditional Marginal Expected Shortfall (MES) framework of Acharya et al. (2017) and Brownlees and Engle (2017) is adapted to the case of energy price shocks. In a first step, the expected return of each energy asset is estimated conditional on an energy market crisis. The estimation combines a vector-error correction model (VECM) and a multiplicative GARCH specification, capturing Granger causality in both the mean and the variance of returns. A dynamic principal components analysis is used to extract latent common factors and decompose standardized residuals into factor-driven and idiosyncratic components. Tail expectations are then estimated nonparametrically, based on these latent factors.

M. Taniguchi et al., *Econometrics, Finance, and Time Series Analysis*,
JSS Research Series in Statistics,
https://doi.org/10.1007/978-981-95-8045-3_4

In a second step, the estimated exposures are combined with asset-specific prices and quantity exposures to derive the total monetary cost of a given energy asset during an energy crisis. Quantity exposures are based on expected final consumption, adjusted for available reserves and contract delivery structures.

Finally, the broader macro-financial implications are assessed by estimating the conditional return of the non-energy equity index under energy stress, relative to normal times. This permits an evaluation of how the impact of energy shocks on the non-energy sector varies with energy market conditions.

The analysis is based on daily futures prices for electricity, natural gas, coal, and EU emission allowances traded on the European Energy Exchange (EEX), along with equity returns for industrial firms in the DAX index. The methodology is applied to several historical stress episodes, including the 2009 Russia–Ukraine gas dispute, the 2010 BP oil spill, and the 2011 Fukushima disaster. The framework provides a tool for decomposing aggregate risk into asset-level contributions and for quantifying the macroeconomic transmission of energy shocks through financial markets.

4.2 Risk Measures and Estimation

4.2.1 Risk Measures Definition

This section defines the three core measures used to assess the exposure of the non-energy sector to extreme energy price movements. All expectations are conditional on information available at time $t-1$. The objective is to model the response of asset prices to an energy crisis—defined as a sharp and concurrent rise in energy prices and deterioration in non-energy equity prices.

The *conditional Marginal Expected Shortfall* (MES) of energy asset i is:

$$MES_{it}(C) = \mathrm{E}_{t-1}\left(r_{it} | r_{EnM,t} > C,\ r_{M,t} < 0\right), \tag{4.1}$$

where r_{it} denotes the return of asset i, $r_{EnM,t}$ is the return of the energy market factor, $r_{M,t}$ is the return of the industrial (non-energy) sector. The threshold C is set as a high quantile (e.g., 95th percentile) of the unconditional distribution of the energy market return. This formulation captures the average loss of energy asset i in the event of an energy price spike coinciding with a downturn in the non-energy market.

The *Energy Systemic Risk* (EnSysRISK) measure aggregates expected price declines with physical exposure to the asset:

$$EnSysRISK_{it} = \max(0,\ p_{it-1} * \exp(MES_{it}) * w_{it}), \tag{4.2}$$

where p_{it-1} is the previous price of energy asset i, and w_{it} is the quantity exposure of the economy to energy asset i at time t.[1] Thus, the EnSysRISK measure captures the total monetary cost to the economy stemming from its exposure to asset i during an energy crisis.

Finally, the *net marginal impact* of an energy crisis on the non-energy sector (ΔMES) is defined as:

$$\Delta MES_{Mt}(C) = MES_{Mt}(C) - \mathrm{E}_{t-1}\left(r_{Mt}|r_{EnMt} > q_{0.5}, r_{Mt} < 0\right) \tag{4.3}$$

where $MES_{Mt}(C)$ is the MES of the industrial sector under an energy crisis, and the second term is the expected return under a median-level energy market return (i.e., $q_{0.5}$ is the median of the energy return sample distribution). This measure is inspired by the $\Delta CoVaR$ of Adrian and Brunnermeier (2016) and represents the difference between the expected return of the non-energy index during an energy crisis and the expected return during 'normal' times. The energy crisis is expected to have a negative net impact on the economy (i.e., ΔMES is expected to be negative).

4.2.2 Econometric Methodology

The estimation of MES relies on decomposing the conditional expectation into its constituent components: conditional mean, conditional volatility, and tail expectation of standardized residuals. Formally,

$$\begin{aligned} MES_{it}(C) &= \mathrm{E}_{t-1}\left(\mu_{it} + \sigma_{it} u_{it}|r_{EnMt} > C,\ r_{Mt} < 0\right) \\ &= \mu_{it} + \sigma_{it}\mathrm{E}_{t-1}\left(u_{it}|r_{EnMt} > C,\ r_{Mt} < 0\right) \end{aligned} \tag{4.4}$$

where μ_{it} and σ_{it}^2 are the conditional mean and variance of asset return i and $u_{it} = (r_{it} - \mu_{it})/\sigma_{it}$ are the standardized residuals. The modeling of μ_{it}, σ_{it}, and u_{it} is carried out using three components: a cointegrated mean model, a multiplicative GARCH structure, and a dynamic factor model for tail expectations.

This methodology relates to the literature on common movements and the modeling of the joint distribution of energy prices. Cointegration was found in the spot prices of different European regional markets (Escribano et al. 2011; Haldrup and Nielsen 2006), in natural gas and electricity futures prices (Emery and Liu 2001), and

[1] For an energy contract i with maturity τ_{i0} and delivery period ν, the exposure at time t is

$$w_{it} = \frac{\varsigma_i}{\nu_i} \sum_{\tau=\tau_{i0}}^{\tau_{i0}+\nu_i} \mathrm{E}_{t-1}\left(fincons_\tau - inv_\tau\right) \quad \text{with } t < \tau_{i0} \leq \tau,$$

where $fincons_\tau$ is the daily final consumption of energy during delivery period ν_i of contract i starting at τ_{i0}, inv_τ are the energy reserves available to the non-energy sector during the same delivery period, and ς_i is the proportion of energy delivered during period ν_i via energy futures contracts i.

in electricity futures of different maturities (Bauwens et al. 2013). Bunn and Fezzi (2008) model electricity, gas and carbon returns using a vector-error correction model with one cointegration vector. Bauwens et al. (2013) propose a semi-parametric multiplicative DCC model for the multivariate volatility of electricity futures contracts of different maturities. Chevallier (2012) find cross-volatility spillovers and time-varying correlations in oil, gas and carbon returns applying different multivariate GARCH models. Benth and Kettler (2010) propose a dynamic copula to model the joint distribution of electricity and gas prices. Copulas are also used by Boerger et al. (2009) and Gronwald et al. (2011) to model the dependence of different energy commodities.

4.2.2.1 Granger Causality in Mean and Variance

Following the methodology of Billio et al. (2012), we apply Granger causality tests to measure the degree of interconnectedness of the energy market. The causal relationships reflect physical relationships in the energy market based on the supply curve (merit-order) of electricity and substitutions between primary energy commodities for electricity generation and other consumption purposes. Causal relationships also reflect financial relationships through the term structure of futures prices and possible spillover effects between energy and industrial markets. We test for causal relationships in returns means and variances using an augmented vector-error correction model for the means and a multiplicative causality GARCH model for the variances.

Augmented Vector-Error Correction Model

To capture dynamic interactions, cointegration, and seasonal components, the conditional mean of each return r_{it} is specified via:

$$r_{it} = \pi_i \eta' \ln(\mathrm{p}_{t-1}) + \sum_{k=1}^{K} \delta'_{ik} \mathrm{r}_{t-k} + \sum_{m=1}^{M} \theta'_{im} \mathrm{x}_{t-m} + \varphi'_i \mathrm{q}_t + \epsilon_{it} \tag{4.5}$$

where η are the cointegrating vectors (estimated via Johansen's trace test), π_i are error correction parameters, δ_{ik} is a $(n \times 1)$ vector of autocorrelation and Granger-causal parameters of order k, x_{t-m} include exogenous variables lagged by m days and q_t are deterministic (seasonal) indicators. This VECM captures both short-run autocorrelations and long-run equilibrium relationships among energy prices, allowing for multi-directional causal dynamics.

Multiplicative GARCH models

The conditional variance of residuals is modeled through a multiplicative structure:

$$\epsilon_{it} = \sigma_{it} u_{it} = \sqrt{\phi_{it} g_{it}} u_{it} \tag{4.6}$$

where

$$g_{it} = (1 - \alpha_{ii} - \beta_i - \frac{\gamma_{ii}}{2}) + \alpha_{ii}\left(\frac{\epsilon_{it-1}^2}{\phi_{it-1}}\right) + \beta_i g_{it-1} + \gamma_{ii}\left(\frac{\epsilon_{it-1}^2}{\phi_{it-1}}\right) I_{\{\epsilon_{it-1}<0\}}, \quad (4.7)$$

$$\phi_{it} = f\left(u_{1t-1}, ..., u_{i-1,t-1}, u_{i+1,t-1}, ..., u_{nt-1}\right) l_i(t), \quad (4.8)$$

$I_{\{\epsilon_{it-1}<0\}}$ is a dummy variable equal to one when the past shock of asset i is negative, and $l_i(t)$ is a deterministic function of time. Here, g_{it} captures own volatility persistence (augmented GJR-GARCH), and ϕ_{it} reflects volatility spillovers from other assets. This multiplicative structure ensures interpretability and tractable joint estimation.[2]

4.2.2.2 Dynamic PCA and Tail Expectation Estimation

The standardized residuals may be decomposed as a linear function of common factors $\mathrm{y}_t = (y_{1t}, y_{2t}, ..., y_{st})$ and idiosyncratic terms ζ_{it}

$$u_{it} = \sum_{j=1}^{s} a_{ijt} y_{jt} + \zeta_{it} \quad (4.9)$$

where y_{jt} are obtained from a dynamic principal component analysis, a_{ijt} is the element of the eigenvector associated with asset i and principal component y_{jt}, $\zeta_{it} = u_{it} - \sum_{j=1}^{s} a_{ijt} y_{jt}$, and $s \leq n$.

Dynamic PCA

The PCA is applied to the dynamic correlation matrix R_t, which is computed from a Dynamic Conditional Correlation (DCC) model (Engle 2002). The DCC process is defined for the $n \times n$ symmetric positive-definite matrix Q_t by

[2] Note that, if $\epsilon_{it-1}^2/\phi_{it-1}$ are replaced by $\epsilon_{it-1}^2/l_i(t-1)$ in (4.7), then a model similar to the exponential causality GARCH model of Caporin (2007) would be obtained, with additional deterministic factors $l(t)$. The advantage of standardizing returns by ϕ_{it} rather than $l_i(t)$ is that the process g_{it} becomes a standard GARCH/GJR process, for which theoretical results are broadly available. This model can also be viewed as a simplified version of the spline-GARCH model of Engle and Rangel (2008) but the components in the multiplicative causality model are both low frequency components. When the second component simplifies to a constant, the equation simplifies to the GARCH or GJR model.

In our application, the interaction component is specified as

$$\phi_{it} = c_i \exp\left(\sum_{j=1, j\neq i}^{n} \left(\vartheta_{ij} u_{jt-1} + \alpha_{ij}|u_{jt-1}|\right) + \kappa_i' \mathrm{d}_t\right),$$

where d_t are deterministic terms including seasonal dummies. This function has a similar form as the EGARCH model of Nelson (1991) and allows for asymmetric causality effects when $\vartheta_{ij} \neq 0$.

$$Q_t = (1 - a - b)\bar{Q} + au_{t-1}u'_{t-1} + bQ_{t-1} \tag{4.10}$$

where $a + b < 1, a, b \geq 0$ and $\bar{Q}$ is a parameter matrix. Then the DCC correlation matrix is obtained by transforming this to

$$R_t = (\text{diag}Q_t)^{-1/2} Q_t (\text{diag}Q_t)^{-1/2}$$

and the covariance matrix of mean zero residuals ($\epsilon_{it} = r_{it} - \mu_{it}$) is given by $H_t = D_t R_t D_t$ where $D_t = \text{diag}(\sigma_{1t}, \sigma_{2t}, ..., \sigma_{nt})$ is a $n \times n$ diagonal matrix collecting the univariate conditional volatilities of residuals on the diagonal.[3]

From the spectral decomposition of the DCC correlation matrix

$$H_t = D_t R_t D_t = D_t \left(A_t \Lambda_t A'_t + R_{\zeta_t}\right) D_t \tag{4.11}$$

where A_t is a matrix of s eigenvectors associated with the s largest eigenvalues that are contained in the diagonal matrix $\Lambda_t = \text{diag}(\lambda_{1t}, \lambda_{2t}, ..., \lambda_{st})$ with $\lambda_{1t} \geq \lambda_{2t} \geq ... \geq \lambda_{st}$, and R_{ζ_t} is the correlation matrix of idiosyncratic terms ζ_t.

To isolate the energy factor and ensure orthogonality with the industrial (non-energy) sector, we impose a restriction on the first component y_{1t}. Specifically, the non-energy index return r_{Mt} is imposed as the first principal component:

$$\max_{a_{1t}} a'_{1t} R_t a_{1t}$$

$$\text{s.t. } a'_{1t} a_{1t} = 1, \ a_{i1t} = 0 \quad \forall t, \forall i \neq M$$

forcing all loadings for energy assets to be zero in this component. Subsequent components y_{jt} for $j \geq 2$ are extracted using standard PCA subject to orthogonality constraints:

$$\max_{a_{jt}} a'_{jt} R_t a_{jt}$$

$$\text{s.t. } a'_{jt} a_{jt} = 1, \ a'_{jt} a_{lt} = 0 \quad \forall t, \forall l \neq j$$

The second component y_{2t}, denoted y_{EnMt}, captures the common dynamics of energy returns and serves as the proxy for the energy market factor.

Tail Expectations

The conditional probability of an energy crisis $\text{P}_{t-1}(r_{EnMt} > C, \ r_{Mt} < 0)$ is approximated by

[3] If the distribution of $z_t = H_t^{1/2}\epsilon_t$ in (4.10) is assumed Gaussian, the DCC model is estimated by a two-stage quasi-maximum likelihood (QML) estimation procedure, where in the first stage the parameters of the conditional variance processes are estimated. In the second stage, the parameters of the conditional correlation process are estimated conditionally on the estimates obtained in the first stage.

$$\mathrm{P}_{t-1}(y_{EnMt} > C,\ y_{Mt} < 0) = \mathrm{P}(y_{EnMt}/\sqrt{\lambda_{EnMt}} > C/\sqrt{\lambda_{EnMt}},\ y_{Mt} < 0).$$

Tail expectations are then approximated by

$$\begin{aligned}\mathrm{E}_{t-1}\left(u_{it}|r_{EnMt} > C,\ r_{Mt} < 0\right) \simeq \textstyle\sum_{j=1}^{s}\left[a_{ijt}\mathrm{E}_{t-1}\left(y_{jt}|y_{EnMt} > C,\ y_{Mt} < 0\right)\right] \\ +\mathrm{E}_{t-1}\left(\zeta_{it}|y_{EnMt} > C,\ y_{Mt} < 0\right).\end{aligned} \tag{4.12}$$

These conditional expectations are estimated nonparametrically using a kernel estimator[4]:

$$\hat{\mathrm{E}}\left(\zeta_{it}|y_{EnMt} > C,\ y_{Mt} < 0\right) = \frac{\sum_{\tau=1}^{T}\zeta_{i\tau}\Phi\left[\left(\frac{y_{EnM\tau}}{\sqrt{\lambda_{EnM\tau}}} - \frac{C}{\sqrt{\lambda_{EnMt}}}\right)h^{-1}\right]I\left(y_{M\tau} < 0\right)}{\sum_{\tau=1}^{T}\Phi\left[\left(\frac{y_{EnM\tau}}{\sqrt{\lambda_{EnM\tau}}} - \frac{C}{\sqrt{\lambda_{EnMt}}}\right)h^{-1}\right]I\left(y_{M\tau} < 0\right)} \tag{4.13}$$

where $\Phi(\cdot)$ is the Gaussian cumulative distribution function, h is a positive bandwidth, and $I\left(y_{M\tau} < 0\right)$ is an indicator function equal to one when the non-energy factor is negative. The same estimation procedure applies to $\mathrm{E}\left(y_{jt}|y_{EnMt} > C,\ y_{Mt} < 0\right)$. Observations are weighted based on proximity to the threshold C, scaled by the time-varying volatility λ_{EnMt}, ensuring sensitivity to market stress conditions.

This full decomposition and estimation procedure allows the MES to respond flexibly to evolving market conditions, capturing both tail dependence and time-varying volatility.

4.3 Empirical Illustration

4.3.1 Data Description

The analysis relies on daily price series for fourteen financial instruments: ten energy futures, three spot indices, and the DAX industrial equity index representing the non-energy sector. The portfolio of energy products consists of Brent crude oil, European coal, and European Union carbon emission Allowances (EUA) spot indices, together with futures contracts on electricity, natural gas, coal, and EU emission allowances traded on the European Energy Exchange (EEX).

Electricity futures are financial futures written on the German Physical Electricity Index; gas futures are physical futures for the German Gaspool market area; and coal futures are financial futures written on the API#2 index for the ARA region. For EUA futures, the underlying is the delivery of EUA for the second EU Emission Trading Scheme (ETS) period.

---

[4] Nonparametric estimation of tail expectations is an alternative to copula functions where the joint distribution of idiosyncratic and factor disturbances is left unspecified. Brownlees and Engle (2017) propose a kernel estimator of tail expectations based on the literature on the nonparametric estimation of the expected shortfall (Scaillet 2004) and the conditional expected shortfall (Scaillet 2005; Kato 2012).

The futures contracts have monthly, quarterly, and yearly maturities and corresponding delivery periods, except for EUA futures, which have yearly maturity and delivery during the second EU ETS period (a five-year phase starting on January 1, 2008). The futures price series are constructed from successive nearest contracts over the period 07.03.2007 to 06.01.2011, and returns are adjusted for contract switches.

The dataset also includes the Brent crude oil price per barrel and the Merrill Lynch Commodity Index (MLCX) for EUA and European coal spot markets. The DAX industrial index, composed primarily of energy-consuming companies, serves as a proxy for the non-energy sector in the analysis.

4.3.2 Results

The conditional MES of energy assets is estimated using the methodology described in Sect. 4.2. Figure 4.1 displays the cross-sectional average of the conditional MES for each energy commodity class.

Across all commodities, MES rises significantly during and after the 2008 financial crisis, with a peak in early January 2009 coinciding with high volatility, a cold winter, and the disruption in European gas supply caused by the Russia–Ukraine gas dispute (January 2009). Subsequent spikes in MES are observed in response to

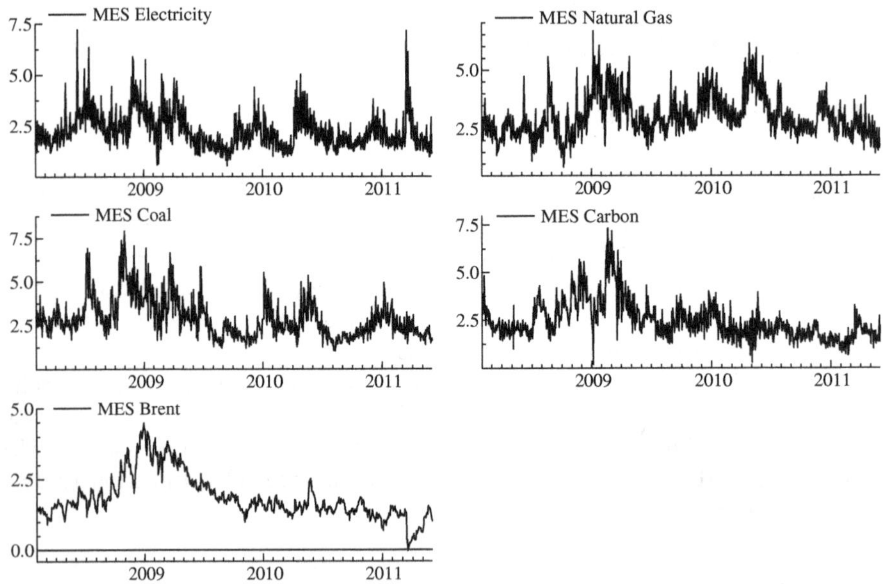

Fig. 4.1 The average conditional MES ($MES_{it}(C)$) in percentage for each energy commodity class (Electricity, Natural Gas, Coal, EUA, Crude oil). C is the unconditional VaR at 95% of the market factor

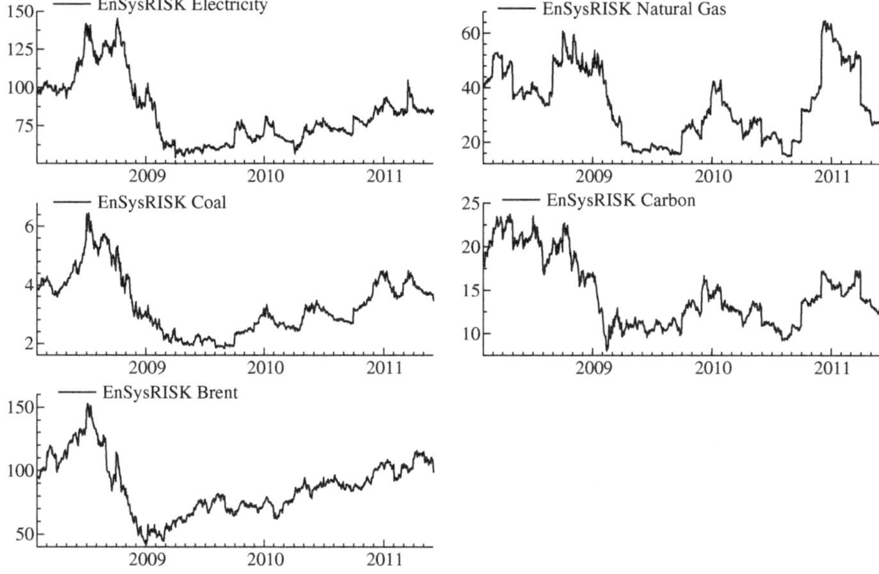

Fig. 4.2 Energy Systemic Risk measure (EnSysRISK) in million euros for each energy class (Electricity, Natural Gas, Coal, EUA, Crude oil)

the BP Deepwater Horizon oil spill (April 2010) and the Fukushima nuclear disaster (March 2011), particularly for natural gas and electricity futures. In Germany, the political decision to phase out nuclear power after Fukushima results in a sharp increase in the MES of electricity futures.

The EnSysRISK measure combines each asset's conditional MES with its price and estimated quantity exposure to the non-energy sector.[5] The result is a forward-looking estimate of the total monetary cost to the economy during an energy crisis, shown in Fig. 4.2.

EnSysRISK exhibits a rising trend over the sample, driven by increases in prices and final consumption. For natural gas, the EnSysRISK measure shows a pronounced seasonal pattern, peaking in winter due to its use in heating. The measure is particularly elevated in winter 2009 (Russia–Ukraine gas dispute) and again in winter 2011. Electricity total costs, as measured by EnSysRISK, also increase after Fukushima, though the effect dissipates by May 2011. In general, EnSysRISK is largest of electricity, crude oil, and natural gas, reflecting the higher total costs of these commodities during an energy crisis.

Finally, Fig. 4.3 shows the estimated ΔMES for the DAX industrial index, representing the net impact of an energy crisis on the non-energy sector. The measure

[5] The quantity exposure is proxied by final consumption, implying conservative estimates for storable commodities. Monthly energy consumption data are downloaded from Eurostat (natural gas, coal, crude oil) and Entso-e (electricity), and converted in the adequate quantity units (corresponding to the price definitions) using the conversion factors of the BP Statistical Review (2011).

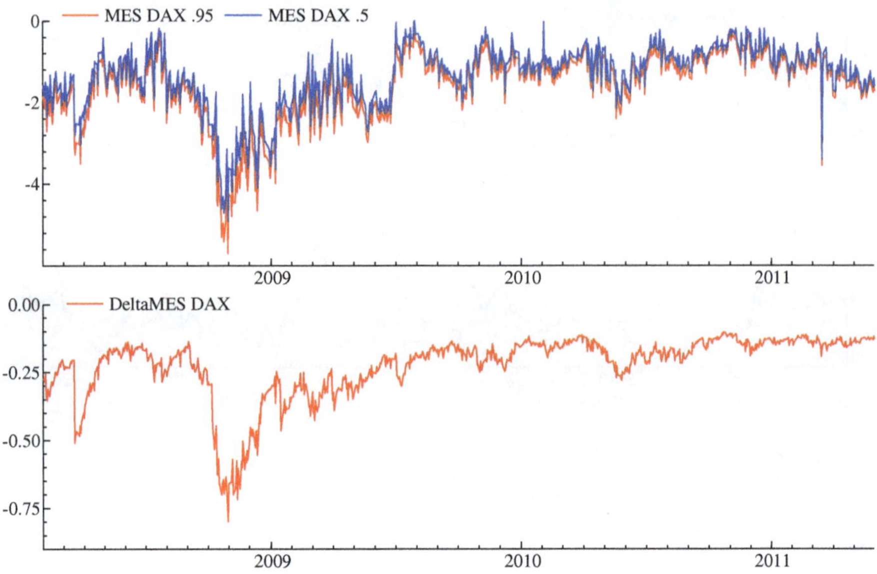

Fig. 4.3 Impact of energy crises on the economy: $MES(.95)$ and $MES(.5)$ (upper panel), and ΔMES (lower panel) of the DAX industrial index

is negative throughout the sample, confirming that extreme energy price increases tend to depress industrial equity prices. However, the magnitude of ΔMES varies. The largest decline is observed during the height of the financial crisis (September 2008 to January 2009), suggesting that energy shocks exert a greater impact when macroeconomic conditions are already weak.

4.4 Conclusion

This chapter develops an econometric framework to quantify the exposure of the non-energy sector to extreme energy price shocks. By adapting the conditional Marginal Expected Shortfall methodology to energy markets, the analysis links extreme increases in energy prices to expected losses in industrial equity returns. The approach combines cointegration, volatility spillovers, and tail dependence with a latent energy market factor, to capture the complex dependence structure of energy assets during stress episodes.

The empirical results indicate that energy price shocks generate economically meaningful losses for the non-energy sector, particularly during periods of heightened macroeconomic fragility. Electricity, crude oil and natural gas emerge as the main contributors to aggregate risk, reflecting both their volatility and their importance for final consumption. While energy market disruptions do not uniformly translate into large equity losses, their impact is markedly stronger during downturns, as illustrated by the financial crisis and major energy supply shocks.

Overall, the framework provides a flexible and forward-looking tool to assess the transmission of energy market stress to the real economy. By decomposing aggregate exposure into asset-specific contributions, it offers a basis for evaluating how changes in the energy mix, market integration, or policy interventions may alter the vulnerability of the non-energy sector to future energy price shocks.

References

Acharya, Viral V et al. (2017). "Measuring systemic risk". In: *The Review of Financial Studies* 30.1, pp. 2–47.

Adrian, Tobias and Markus K Brunnermeier (2016). "CoVaR". In: *The American Economic Review* 106.7, pp. 1705–1741.

Bauwens, L., C. Hafner, and D. Pierret (2013). "Multivariate Volatility Modeling of Electricity Futures". In: *Journal of Applied Econometrics* 28:5, pp. 743–761.

Benth, F. E. and P. Kettler (2010). "Dynamic copula models for the spark spread". In: *Quantitative Finance* 11:3, pp. 407–421.

Billio, M. et al. (2012). "Econometric measures of connectedness and systemic risk in the finance and insurance sectors". In: *Journal of Financial Economics* 104, pp. 535–559.

Boerger, R. H. et al. (2009). "Cross-Commodity Analysis and Applications to Risk Management". In: *Journal of Futures Markets* 29:3, pp. 197–217.

Brownlees, Christian and R.F. Engle (2017). "SRISK: A conditional capital shortfall measure of systemic risk". In: *The Review of Financial Studies* 30.1, pp. 48–79.

Bunn, D. and C. Fezzi (2008). "A vector error correction model of the interactions among gas, electricity and carbon prices: An application to the cases of Germany and United Kingdom". In: *Markets for Carbon and Power Pricing in Europe: Theoretical Issues and Empirical Analyses*. Francesco Gulli, pp. 145–159.

Caporin, M. (2007). "Variance (Non) Causality in Multivariate GARCH". In: *Econometric Reviews* 26:1, pp. 1–24.

Chevallier, J. (2012). "Time-varying correlations in oil, gas and CO2 prices: an application using BEKK, CCC and DCC-MGARCH models". In: *Applied Economics* 44:32, pp. 4257–4274.

Emery, G. and Q. Liu (2001). "An analysis of the relationship between electricity and natural gas futures prices". In: *Journal of Futures Markets* 22, pp. 95–122.

Engle, R.F. (2002). "Dynamic conditional correlation: a simple class of multivariate generalized autoregressive conditional heteroskedasticity models". In: *Journal of Business and Economic Statistics* 20:3, pp. 339–350.

Engle, R.F. and J.G. Rangel (2008). "The spline-GARCH model for low-frequency volatility and its global macroeconomic causes". In: *Review of Financial Studies* 21:3, pp. 1187–1222.

Escribano, A.J., I. Pena, and P. Villaplana (2011). "Modeling Electricity Prices: International Evidence". In: *Oxford Bulletin of Economics and Statistics* 73:5, pp. 622–650.

Gronwald, M., J. Ketterer, and S. Trck (2011). "The relationship between carbon, commodity and financial markets - a copula analysis". In: *Economic Record* 87, pp. 105–124.

Haldrup, N. and M. Nielsen (2006). "A regime switching long memory model for electricity prices". In: *Journal of Econometrics* 135:1-2, pp. 349–376.

Kato, K. (2012). "Weighted Nadaraya-Watson Estimation of Conditional Expected Shortfall". In: *Journal of Financial Econometrics* 10:2, pp. 265–291.

Nelson, D. B. (1991). "Conditional Heteroskedasticity in Asset Returns: a New Approach". In: *Econometrica* 59, pp. 349–370.

Scaillet, O. (2004). "Nonparametric estimation and sensitivity analysis of expected shortfall". In: *Mathematical Finance* 14, pp. 115–129.
Scaillet, O. (2005). "Nonparametric estimation of conditional expected shortfall". In: *Insurance and Risk Management Journal* 74, pp. 639–660.

Chapter 5
Modeling Solvency–Liquidity Interactions in Banking: A Panel VAR Analysis

Abstract This chapter analyzes the dynamic interaction between solvency and liquidity risks in banking using a panel VAR framework. Banks with higher expected capital shortfalls reduce short-term funding, while those with greater reliance on short-term debt experience increases in future solvency risk. These feedback effects are persistent, asymmetric, and more pronounced for capital-constrained banks. Impulse response functions trace the propagation of solvency and liquidity shocks through banks' balance sheets and reveal significant heterogeneity in adjustment dynamics across institutions and over time.

Keywords Capital shortfall · Funding liquidity risk · Short-term funding · Solvency–liquidity interaction

5.1 Introduction

Banks transform short-term funding into long-term investments, exposing them to both liquidity and solvency risks. A rich theoretical literature argues that solvency and liquidity are mutually reinforcing: a deterioration in fundamentals triggers creditor withdrawals, and the ensuing loss of funding forces asset fire-sales that further erode capital (Allen and Gale 1998; Diamond and Rajan 2005; Morris and Shin 2008). In spite of these insights, post-crisis rule-making continues to treat the two risks in isolation—most visibly in the Basel III capital regime and its Liquidity Coverage Ratio (Basel Committee on Banking Supervision 2011, 2013).

To study the interplay of these risks, this chapter relies on SRISK (Acharya et al. 2017; Brownlees and Engle 2017), a forward-looking market-based measure of solvency risk. SRISK estimates the expected capital shortfall a bank would face in the event of a systemic market decline and is constructed using data on market capitalization, leverage, and downside risk exposure. It reflects the amount of capital that would be needed to restore a bank's capital ratio to a regulatory threshold following a severe market shock.

M. Taniguchi et al., *Econometrics, Finance, and Time Series Analysis*,
JSS Research Series in Statistics,
https://doi.org/10.1007/978-981-95-8045-3_5

In this chapter, a fixed-effects panel vector autoregressive (VAR) framework is employed to model banks' short-term balance sheet flows jointly with stressed solvency measures. Empirical evidence for the solvency–liquidity nexus was documented by Pierret (2015); the present chapter broadens that contribution by recasting the analysis in a dynamic panel VAR setting and by quantifying how funding structure and liquidity positions affect forward-looking solvency risk indicators such as SRISK. After estimation, impulse response functions are derived to trace how solvency shocks propagate into funding flows—and how liquidity shocks feed back into expected capital shortfalls—over time.

Three main contributions are presented in this chapter. First, evidence is provided of a solvency–liquidity interaction: banks with higher expected capital shortfalls (higher SRISK) raise less short-term debt, and banks with more short-term debt become more vulnerable to solvency risk. Second, impulse response functions are used to illustrate how solvency and liquidity shocks propagate over time and vary across banks, especially distinguishing between adequately capitalized and capital-constrained institutions. Third, the interaction between profitability and solvency is found to influence short-term balance sheet dynamics, especially for capital-constrained banks.

5.2 Data and Variables

Quarterly regulatory reports filed on Form FR Y-9C and daily market information are merged for a sample of forty-four publicly traded U.S. bank-holding companies over the period 2000 Q1–2013 Q1. Short-term debt and short-term assets are constructed from FR Y-9C items obtained via the SNL Financial database. Short-term debt ($\mathrm{STDebt}_{\mathrm{it}}$) of bank i at time t includes uninsured time deposits with less than one year to maturity, repurchase agreements (repos), federal funds purchased, and other borrowed funds maturing within a year. Short-term assets ($\mathrm{STAssets}_{\mathrm{it}}$) of bank i at time t comprise debt securities maturing within a year, interest-bearing bank balances (cash), reverse repos, and federal funds sold. The difference between short-term debt and assets is labeled the liquid asset shortfall, which proxies a bank's exposure to funding liquidity risk. Profitability is also used in some specifications, defined as the bank net income ($\mathrm{NI}_{\mathrm{it}}$) divided by its total assets ($\mathrm{TA}_{\mathrm{it}}$), and is also drawn from the FR Y-9C reports.

Solvency risk is measured using SRISK, a forward-looking, market-based estimate of the expected capital shortfall of a bank in a systemic event (Acharya et al. 2017; Brownlees and Engle 2017). SRISK is the expected capital shortfall a bank would face if equity markets declined by 40% over a six-month horizon. Formally, SRISK of bank i at time t is defined by

$$\begin{aligned}\mathrm{SRISK}_{\mathrm{it}} &= \mathrm{E}_t[k(\mathrm{D}_{\mathrm{it}+h} + \mathrm{MV}_{\mathrm{it}+h}) - \mathrm{MV}_{\mathrm{it}+h} | R_{\mathrm{mt}+h} \leq -0.4] \\ &= k\mathrm{D}_{\mathrm{it}} - (1-k) * \mathrm{MV}_{\mathrm{it}} * (1 - \mathrm{LRMES}_{\mathrm{it}})\end{aligned} \tag{5.1}$$

where $R_{\mathrm{mt}+h}$ is the return of the market index from period t to period $t+h$ ($h =$ 6 months), k is the regulatory capital ratio (8%), D_{it} is the book value of debt, $\mathrm{MV}_{\mathrm{it}}$ is the market capitalization, $\mathrm{LRMES}_{\mathrm{it}}$ is the long-run marginal expected shortfall (bank's expected return conditional on a large market decline) defined as $\mathrm{LRMES}_{\mathrm{it}} = -\mathrm{E}_t(R_{\mathrm{it}+h} | R_{\mathrm{mt}+h} \leq -0.4)$. SRISK thus combines size, leverage, and market sensitivity into a single crisis-based solvency metric and is available at high frequency from public data. These measures (SRISK and LRMES) are available from the V-Lab website developed at NYU Stern School of Business.[1]

5.3 Methodology: Panel VAR Model

To estimate the dynamic interaction between solvency and liquidity, a fixed-effects panel vector autoregressive (VAR) model is employed. The VAR framework allows modeling the joint evolution of several interdependent time series while accounting for feedback effects. In this context, the model tracks the evolution of short-term debt, short-term assets, and solvency risk across banks and over time.

Let $w_{\mathrm{it}} = (y_{\mathrm{it}}, z_{\mathrm{it}}, \mathrm{SRISK}_{\mathrm{it}}/\mathrm{TA}_{\mathrm{it}})'$ denote the vector of endogenous variables for bank i at time t, where $y_{\mathrm{it}} = \log(\mathrm{STDebt}_{\mathrm{it}})$, $z_{\mathrm{it}} = \log(\mathrm{STAssets}_{\mathrm{it}})$, $\mathrm{SRISK}_{\mathrm{it}}$ is defined in Eq. (5.1), and $\mathrm{TA}_{\mathrm{it}}$ are the bank total assets. The baseline panel VAR model for the $(K \times 1)$ vector of endogenous variables w_{it} is specified as:

$$w_{\mathrm{it}} = \alpha_i + \Phi w_{\mathrm{it}-1} + \theta t + \varepsilon_{\mathrm{it}}, \qquad t = 1, 2, ..., T_i,\ i = 1, 2, ..., N, \tag{5.2}$$

where α_i denotes bank-specific fixed-effects, θ is a trend parameter, Φ is a $(K \times K)$ matrix of VAR parameters, and $\varepsilon_{\mathrm{it}}$ is a $(K \times 1)$ vector of error terms. Panel unit root tests confirm that the variables are trend-stationary, which permits estimation in levels. The panel dataset is unbalanced but we require $T_i \geq 30$. The model is estimated via pooled ordinary least squares with bank fixed-effects.[2]

To allow for heterogeneous dynamics and based on in-sample fit criteria (Wald test and adjusted R^2), an extended model specification includes bank-specific autoregressive and trend coefficients:

$$w_{\mathrm{it}} = \alpha_i + \phi_i \odot w_{\mathrm{it}-1} + \theta_i t + \delta w_{\mathrm{it}-1} + \varepsilon_{\mathrm{it}}, \tag{5.3}$$

where ϕ_i, θ_i are $(K \times 1)$ vectors of parameters specific to bank i, δ is a $(K \times K)$ matrix of parameters with zeros on the diagonal, and $\odot$ is the Hadamard product.

[1] See http://vlab.stern.nyu.edu/.

[2] The parameters of Eq. (5.2) are estimated by ordinary least squares. Judson and Owen (1999) report severe negative bias for the autoregressive parameters of dynamic panel regressions due to the small time series dimension even when $T = 30$. The potential negative bias of autoregressive parameters has implications for the stationarity of the endogenous variables of the panel VAR. Running the regressions in first differences does not, however, qualitatively change the results on the δ parameter estimates.

Bank-specific parameters mainly reflect different business models and the resulting differences in aversion for funding liquidity risk and solvency risk.

5.4 Empirical Results

5.4.1 Solvency–Liquidity Nexus

Table 5.1 reports the estimates of the interaction parameters (δ) of Eq. (5.3), where $w_{it} = (y_{it}, z_{it}, \text{SRISK}_{it}/\text{TA}_{it})'$.[3] A statistically significant and negative coefficient of −0.011 is estimated on the ratio of lagged SRISK to total assets (SRISK/TA) in the short-term debt equation. This implies that banks with higher expected capital shortfalls reduce their reliance on short-term funding in the following quarter. Conversely, the lag of short-term debt enters positively in the SRISK equation, with a coefficient of 0.009, indicating that a greater reliance on short-term debt raises the future vulnerability of the bank to systemic stress.

Importantly, the interaction is asymmetric. Solvency risk significantly reduces short-term funding access, while more short-term debt increases the risk of insolvency in a crisis. This finding is consistent with theoretical models that highlight the amplification between fundamentals and funding withdrawals during systemic episodes (Allen and Gale 1998; Diamond and Rajan 2005; Morris and Shin 2008).

The dynamics of short-term assets are weaker: neither SRISK nor short-term debt significantly predict short-term asset holdings in the baseline specification. This suggests that liquid assets are not actively adjusted in response to changing solvency or funding conditions at the quarterly frequency, a result that aligns with the view that banks tend to hoard liquidity under stress rather than rebalance their asset structure rapidly.

5.4.2 Interaction Between Solvency and Profitability

Incorporating profitability allows distinguishing between demand- and supply-driven variations in short-term funding. When banks face profitable investment opportunities, they are likely to demand more short-term funding to finance these activities. At the same time, their liquidity risk increases as funds are deployed into longer-term assets. Following Perotti and Suarez (2011), net income scaled by total assets is used as a proxy for profitability and included as an additional variable in the panel VAR.

To account for the role of profitability and its interaction with capital adequacy, the panel VAR is augmented with a state-dependent specification:

[3] Note that replacing the logarithm of short-term debt and the logarithm of short-term assets (resp. y_{it} and z_{it}) by their ratios to total assets (resp. $\text{STDebt}_{it}/\text{TA}_{it}$ and $\text{STAssets}_{it}/\text{TA}_{it}$) in vector w_{it} does not qualitatively change the results of this chapter.

$$w_{it} = \alpha_i + \phi_i \odot w_{it-1} + \theta_i t + \delta w_{it-1} + \gamma w_{it-1} * s_{t-1} + \omega s_{t-1} + \varepsilon_{it} \qquad (5.4)$$

where ω is a $(K \times 1)$ vector of parameters, γ is a $(K \times K)$ matrix of parameters with zeros on the diagonal. The vector $w_{it} = (y_{it}, z_{it}, \text{SRISK}_{it}/\text{TA}_{it}, \text{NI}_{it}/\text{TA}_{it})'$ is now extended to include four variables: short-term debt, short-term assets, SRISK/TA, and profitability, where profitability is measured as net income divided by total assets. The state variable s_t could be a bank characteristic or a common factor. In Table 5.2 (Panel B), the state variable is an indicator variable equal to one when SRISK is positive ($s_{it} = 1_{\{\text{SRISK}_{it}>0\}}$), i.e., when the bank is expected to have a capital shortfall in a crisis.

The results from this extended specification, reported in Table 5.2, confirm the asymmetric role of profitability. When SRISK is non-positive ($s_{it} = 0$), profitability is associated with greater short-term debt issuance and lower short-term asset holdings, suggesting demand-driven behavior. In contrast, when SRISK is positive ($s_{it} = 1$), the interaction terms offset these effects and the influence of profitability on the balance sheet vanishes. This state-dependent pattern underscores that profitability enhances balance sheet flexibility only when the bank is adequately capitalized.

These results reinforce the interpretation of SRISK as a supply-side constraint on funding. Profitable banks can raise funding in normal conditions, but in states of expected capital shortfall, solvency considerations dominate and restrict access to short-term financing, regardless of profitability.

5.4.3 *Impulse Response Functions*

The dynamic implications of solvency and liquidity shocks are examined through impulse response functions (IRFs) derived from the estimated panel VAR model with profitability. IRFs trace the average time profile of the endogenous variables following a unit shock to one of the system's innovations, holding other shocks constant. In the present context, IRFs reveal how a shock to solvency (SRISK) or liquidity (short-term debt or short-term assets) propagates through the balance sheet over time.

Identification of structural shocks requires an ordering of variables. A Cholesky decomposition is adopted with the following recursive structure: profitability, SRISK, short-term debt, short-term assets. This ordering reflects the assumption that exogenous shocks first impact a firm via its activities and is translated into the net income, markets react by adjusting their assessment of the capital shortfall, this in turn affects how much funding the bank can access, and the short-term assets adjust in consequence.

Due to heterogeneous autoregressive coefficients across banks, as specified in Eq. (5.3), responses to shocks differ at the firm level. Figure 5.1 displays the median impulse response functions to orthogonalized shocks, bounded by the 25th and 75th percentiles across the sample. This range illustrates the heterogeneity of reactions across banks. The responses involving short-term debt exhibit particularly wide dispersion—both in the response of short-term debt to other shocks and in its impact on other variables. The interaction between SRISK and the short-term balance sheet is especially heterogeneous: for some banks, the impact of SRISK shocks on short-term funding dissipates within a few quarters, while for others, the effect persists for up to three years. These patterns underscore the uneven sensitivity of bank funding to solvency stress and demonstrate the persistence of the solvency–liquidity interaction.

The figure illustrates the directional and persistent effects of solvency and liquidity shocks across the system. The IRFs exhibit several key patterns. First, a positive shock to SRISK, interpreted as a deterioration in solvency, leads to a sustained decline in short-term debt over the subsequent quarters. The effect becomes statistically significant in the second quarter and persists for at least one year, with the median response of short-term debt declining by approximately 0.5% relative to its pre-shock level. This response reflects a supply-side tightening of funding conditions for banks that are perceived as vulnerable to systemic stress. Second, a positive innovation to short-term debt increases SRISK in subsequent quarters. The response of SRISK peaks around two quarters after the shock and remains elevated for up to a year. This finding supports the interpretation that excessive short-term funding contributes to future solvency risk, consistent with theories of rollover fragility and maturity mismatch.

The IRFs further support the asymmetric roles played by the key variables. Solvency shocks tend to induce more persistent and pronounced adjustments in short-term debt than the reverse. Additionally, the effect of profitability on short-term funding appears more transient and state-dependent. These findings reinforce the earlier regression results showing that solvency risk acts as a supply constraint in stressed conditions, whereas profitability affects funding when banks are not capital-constrained.

Figure 5.2 presents the gap in median impulse responses between adequately capitalized ($s_{\mathrm{it}} = 0$) and capital-constrained ($s_{\mathrm{it}} = 1$) banks, based on the parameters of Eq. (5.4). The solvency–liquidity nexus is markedly stronger for capital-constrained banks: the response of short-term debt to a SRISK shock is nearly twice as large as that observed in better-capitalized institutions. By contrast, the responses of short-term debt to innovations in profitability or short-term assets are attenuated and fade more quickly. This result highlights that under capital constraints, banks are predominantly governed by solvency-driven supply effects. For these institutions, the solvency–liquidity interaction becomes the dominant determinant of short-term balance sheet dynamics.

Together, Figs. 5.1 and 5.2 provide a rich dynamic perspective on the solvency–liquidity nexus. The impulse response functions highlight the persistent and asymmetric effects of solvency shocks on liquidity, their heterogeneity across institutions, and their amplification under capital constraints. These results reinforce the central message of this chapter: the interplay between solvency and liquidity is both dynamic and conditional on bank-specific fragilities, and forward-looking solvency measures such as SRISK provide valuable insight into systemic vulnerability.

Table 5.1 Testing the solvency–liquidity nexus. This table reports the estimates of the interaction parameters (δ) of Eq. (5.3), with $w_{it} = (y_{it}, z_{it}, \text{SRISK}_{it}/\text{TA}_{it})'$, where $y_{it} = \ln(\text{STDebt}_{it})$, $z_{it} = \ln(\text{STAssets}_{it})$, and $\text{SRISK}_{it}/\text{TA}_{it}$ is the ratio of SRISK to total assets

Dep. variable	y_{it}	z_{it}	$\text{SRISK}_{it}/\text{TA}_{it}$
$\text{SRISK}_{it-1}/TA_{it-1}$	−1.120**	0.074	
	(0.244)	(0.114)	
z_{it-1}	−0.040		−0.001
	(0.023)		(0.002)
y_{it-1}		−0.003	0.009**
		(0.022)	(0.002)
R^2 (%)	20.811	22.157	15.151
Adj. R^2 (%)	15.430	16.868	9.429

Robust standard errors in parentheses. * significant parameter at 5%; ** at 1%. Sample: 2107 panel observations over 2000Q1-2013Q1 (unbalanced), 44 banks

5.5 Conclusion

This chapter investigates the dynamic interplay between solvency and liquidity risk in the banking sector using a fixed-effects panel vector autoregressive (VAR) model. The empirical analysis reveals that solvency and liquidity are tightly interconnected through asymmetric and state-dependent mechanisms. Banks with elevated solvency risk, measured by SRISK, experience constrained access to short-term funding, while excessive reliance on short-term debt increases future vulnerability to systemic shocks. These dynamics are consistent with theoretical predictions of amplification effects between funding fragility and deteriorating fundamentals.

The incorporation of profitability as a bank-specific determinant further highlights the conditional nature of balance sheet adjustments. Profitable banks tend to expand their use of short-term funding and reduce liquid asset holdings when they are adequately capitalized, reflecting demand-driven balance sheet management. However, for undercapitalized banks, this relationship breaks down, and profitability ceases to influence funding outcomes—underscoring the dominance of supply constraints for undercapitalized banks.

Impulse response functions derived from the panel VAR provide additional insight into the propagation of shocks. Solvency shocks produce persistent funding contractions, while increased reliance on short-term funding to delayed increases in expected capital shortfalls. These effects are particularly pronounced for banks identified ex ante as capital-constrained. The findings validate the use of forward-looking market-based solvency indicators such as SRISK and demonstrate the usefulness of dynamic panel methods for analyzing financial fragility.

Overall, the evidence presented in this chapter calls for a more integrated regulatory approach to solvency and liquidity. Treating these risks as mutually reinforcing rather than orthogonal is crucial for anticipating vulnerabilities and designing effective prudential tools—especially under conditions of systemic stress.

Table 5.2 Testing the interaction between solvency and profitability. This table reports the estimates of the interaction parameters (δ) of Eq. (5.4), with $w_{it} = (y_{it}, z_{it}, \text{SRISK}_{it}/\text{TA}_{it}, \text{NI}_{it}/\text{TA}_{it})'$, where $y_{it} = \ln(\text{STDebt}_{it})$, $z_{it} = \ln(\text{STAssets}_{it})$, $\text{SRISK}_{it}/\text{TA}_{it}$ is the ratio of SRISK to total assets, and $\text{NI}_{it}/\text{TA}_{it}$ is the ratio of net income to total assets

	Panel A: No interaction with $s_{it} = 1_{\{\text{SRISK}_{it}>0\}}$				Panel B: Interaction with $s_{it} = 1_{\{\text{SRISK}_{it}>0\}}$			
Dep. variable	y_{it}	z_{it}	$\text{NI}_{it}/\text{TA}_{it}$	$\text{SRISK}_{it}/\text{TA}_{it}$	y_{it}	z_{it}	$\text{NI}_{it}/\text{TA}_{it}$	$\text{SRISK}_{it}/\text{TA}_{it}$
$\text{SRISK}_{it-1}/\text{TA}_{it-1}$	−1.063**	−0.028	−2.225**		−0.935**	−0.120	−1.681**	
	(0.245)	(0.118)	(0.301)		(0.261)	(0.101)	(0.080)	
$\text{SRISK}_{it-1}/\text{TA}_{it-1} * s_{it-1}$					−0.408	1.757*	−5.871**	
					(0.751)	(0.767)	(1.767)	
$\text{NI}_{it-1}/\text{TA}_{it-1}$	2.354	−4.228		−1.217**	9.704**	−7.944*		−2.809**
	(2.278)	(2.331)		(0.389)	(3.290)	(3.716)		(0.935)
$\text{NI}_{it-1}/\text{TA}_{it-1} * s_{it-1}$					−9.902*	6.315		2.134*
					(4.396)	(5.183)		(0.937)
z_{it-1}	−0.038		−0.015	−0.002	−0.033		−0.035	−0.001
	(0.023)		(0.020)	(0.002)	(0.022)		(0.022)	(0.003)
$z_{it-1} * s_{it-1}$					−0.021*		0.078*	−0.0008
					(0.008)		(0.032)	(0.002)
y_{it-1}		−0.004	−0.067**	0.008**		−0.002	−0.031	0.008**
		(0.021)	(0.026)	(0.002)		(0.022)	(0.021)	(0.002)
$y_{it-1} * s_{it-1}$						−0.007*	−0.080*	0.002
						(0.010)	(0.032)	(0.002)
s_{it-1}					0.347*	0.066	0.094	−0.018
					(0.144)	(0.159)	(0.187)	(0.018)
R^2 (%)	20.870	22.318	41.925	15.787	21.278	22.562	44.474	16.184
Adj. R^2 (%)	15.450	16.997	37.977	10.062	15.715	17.089	40.579	10.304

Panel A: model of Eq. (5.4) without state variable (imposing $\gamma = 0$ and $\omega = 0$). Panel B: model of Eq. (5.4) with state variable $s_{it} = 1_{\{\text{SRISK}_{it}>0\}}$. Robust standard errors in parentheses. * significant parameter at 5%; ** at 1%. Sample: 2107 panel observations over 2000Q1-2013Q1 (unbalanced), 44 banks

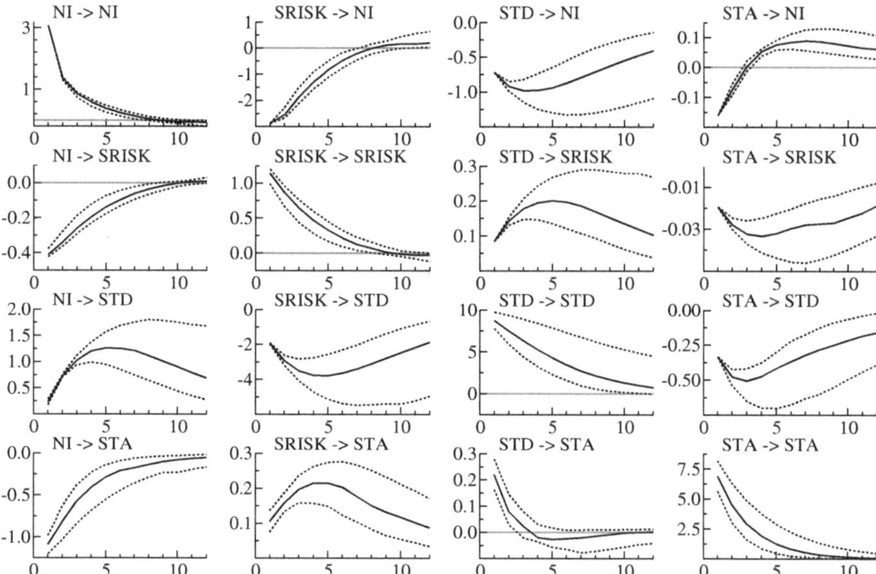

Fig. 5.1 Impulse response functions. Median impulse response function (black lines) between the 25% and 75% impulse response quantiles (dotted lines). "NI" stands for NI_{it}/TA_{it}, "SRISK" for $SRISK_{it}/TA_{it}$, "STD" for $\ln(STDebt_{it})$, and "STA" for $\ln(STAssets_{it})$

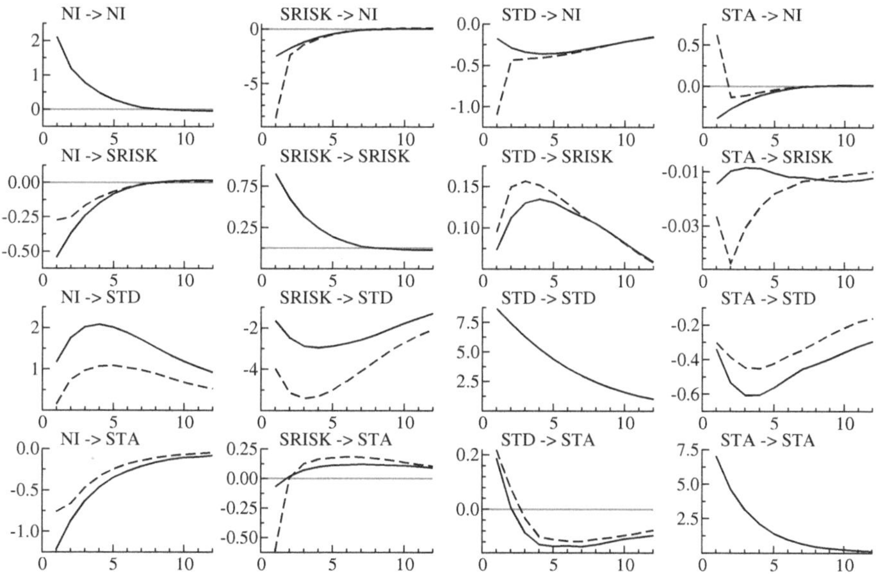

Fig. 5.2 Impulse response function of undercapitalized banks. Impulse response functions with SRISK as a state variable (Eq. (5.4)). Median impulse response function when $SRISK_{it} \leq 0$ (black line) and median impulse response function when $SRISK_{it} > 0$ (dashed line). "NI" stands for NI_{it}/TA_{it}, "SRISK" for $SRISK_{it}/TA_{it}$, "STD" for $\ln(STDebt_{it})$, and "STA" for $\ln(STAssets_{it})$

References

Acharya, Viral V et al. (2017). "Measuring systemic risk". In: *The Review of Financial Studies* 30.1, pp. 2–47.

Allen, F. and D. Gale (1998). "Optimal Financial Crises". In: *Journal of Finance* 53:4, pp. 1246–1284.

Basel Committee on Banking Supervision (June 2011). "Basel III: A global regulatory framework for more resilient banks and banking systems". Bank for International Settlements.

Basel Committee on Banking Supervision (Jan. 2013). "Basel III: The Liquidity Coverage Ratio and Liquidity Risk Monitoring Tools". Bank for International Settlements. URL: http://www.bis.org/publ/bcbs188.pdf.

Brownlees, Christian and R.F. Engle (2017). "SRISK: A conditional capital shortfall measure of systemic risk". In: *The Review of Financial Studies* 30.1, pp. 48–79.

Diamond, D. and R. Rajan (2005). "Liquidity Shortages and Banking Crises". In: *Journal of Finance* 60:2, pp. 615–647.

Judson, R. and A. Owen (1999). "Estimating dynamic panel data models: a guide for macroeconomists". In: *Economics Letters* 65, pp. 9–15.

Morris, S. and H. Shin (2008). "Financial Regulation in a System Context". In: *Brookings Papers on Economic Activity* 2008, pp. 229–261.

Perotti, E. and J. Suarez (2011). "A Pigovian Approach to Liquidity Regulation". In: *International Journal of Central Banking* 27.

Pierret, Diane (2015). "Systemic Risk and the Solvency-Liquidity Nexus of Banks". In: *International Journal of Central Banking* 40.

Chapter 6
Integrated Likelihood-Based Inference for Nonlinear Panel Data Models

Abstract This chapter presents an integrated likelihood approach to the estimation of nonlinear panel data models with individual-specific fixed-effects. Building on the integrated likelihood framework of Severini (2007), the proposed method yields a likelihood that more closely approximates a genuine parametric likelihood than existing approaches in the literature. The statistical properties of the estimator are developed within an asymptotic framework in which both the cross-sectional and time dimensions grow without bound.

Keywords Fixed-effects · Incidental parameters · Integrated likelihood · Likelihood-based inference · Maximum integrated likelihood estimator (MILE) · Nonlinear panel data · Short panels · Zero-score-expectation (ZSE) transformation

6.1 Introduction

In this chapter, we describe a new approach to estimating panel data models in which the individual-specific fixed-effects enter the outcome equation in a nonlinear way. The method is based on the idea of an integrated likelihood due to Severini (2007), but unlike most existing versions discussed in the literature, the one developed here comes much closer to producing what one might call a genuine likelihood function. The accompanying statistical theory is derived in an asymptotic framework where both the number of individuals and the number of time periods grow without bound. Nevertheless, simulation studies show that this integrated likelihood approach performs remarkably well even in panels of moderate size and short duration, both for static and for dynamic models. This chapter is intended to provide an expository overview. The detailed technical results, proofs, and simulation findings are presented in Schumann, Severini, and Tripathi, (2021), hereafter SST21.

M. Taniguchi et al., *Econometrics, Finance, and Time Series Analysis*,
JSS Research Series in Statistics,
https://doi.org/10.1007/978-981-95-8045-3_6

6.2 Panel Data

A panel dataset, also known as a longitudinal dataset, contains observations on a set of individuals followed over time. Such a data structure is particularly valuable in economics and the social sciences, as it allows researchers to estimate and test models of economic and social behavior that include, as explanatory variables, not only characteristics observed by the analyst but also attributes that are unobserved, vary across individuals, and remain constant over time. If these unobserved individual attributes are correlated with some of the observed characteristics, ignoring them can lead to seriously biased statistical inference.

Examples of such time-invariant (at least in the short run) individual characteristics include ability, productivity, or latent cultural preferences—factors that differ across individuals but are typically not observable to the researcher. In the literature, these factors are often referred to as unobserved individual heterogeneity, unobserved individual effects, or simply as "fixed-effects." The latter term is used throughout this chapter and the next to distinguish these unobserved individual attributes from observed time-invariant variables such as gender or ethnicity.

To justify the time-invariance of fixed-effects, panel datasets in microeconometric applications are often characterized as being "short," i.e., having the number of individuals (n) much larger than the number of time periods (T). It is therefore not surprising that the asymptotic theory of inference for microeconometric panel data models is developed under the assumption that $n \to \infty$ and T is held fixed. In this setting, it is well known how to estimate and test models with fixed-effects without making distributional assumptions when the fixed-effects enter the model linearly and additively; cf., e.g., Chamberlain (1982), Hsiao (1986), Baltagi (1995), Arellano (2003b), and Wooldridge (2010). By contrast, parametric distributional assumptions are usually needed to estimate models where the fixed-effects enter nonlinearly. Even then, there is only a limited class of nonlinear models that can be satisfactorily studied in a unified manner due to the well-known "incidental parameters problem" (Lancaster 2000). For instance, while it is known how to consistently estimate a panel logit with fixed-effects (Andersen 1970; Chamberlain 1980), the same approach does not work for estimating probit models, a closely related specification (Magnac 2004; Chamberlain 2010).

In this chapter we describe a unified methodology, based on a new integrated likelihood (IL) approach, for estimating nonlinear panel data models with fixed-effects in a parametric likelihood framework. The new IL can be viewed as a natural extension of the integrated likelihoods introduced by Lancaster (2002) and Arellano and Bonhomme (2009). The statistical theory for the new IL holds in an asymptotic setting in which both n and T increase, with n growing faster than T. Thus, although the approximations underlying the theory are formally justified only in the limit as $n, T \to \infty$, the methodology is nevertheless designed with short panels in mind.

6.3 Likelihood for Panel Data Models

Let Y_{it} denote outcomes and X_{it} a vector of explanatory variables for $i = 1, \ldots, n$ and $t = 1, \ldots, T$. Hereafter, $n, T \geq 2$ and "vector" means a column vector. The random variables Y_{it} and X_{it} are observed, with i indexing the individual and t the time. The fixed-effects α_{i0} are unobserved random variables whose distribution is unknown.

Assumption 6.1 (*Fixed-effects*) For each i, the unobserved random variable α_{i0} is continuously distributed with support $(\mathfrak{a}, \mathfrak{b})$, where $\mathfrak{a}, \mathfrak{b} \in \mathbb{R} \cup \{-\infty, +\infty\}$, $\mathfrak{a} < \mathfrak{b}$, and $\mathfrak{a}, \mathfrak{b}$ are known. □

Henceforth, let $_{iT} := (Y_{i1}, \ldots, Y_{iT})$ denote the time series of outcomes, and $_{iT} := (X_{i1}, \ldots, X_{iT})$ the time series of explanatory variables, corresponding to the ith individual for the duration of the panel.

The distribution of $(_{iT}, \alpha_{i0})$ is unknown, which allows for arbitrary correlation between the fixed-effects and the explanatory variables. Given $(_{iT}, \alpha_{i0})$, the time series $_{iT}$ is drawn from the conditional density $f_{ _{iT}| _{iT},\alpha_{i0};\theta_0}$, which is known up to a parameter $\theta_0 \in \text{int}(\Theta)$, where Θ is a known subset of $\mathbb{R}^{\dim(\theta_0)}$. It is assumed that $f_{ _{iT}| _{iT},\alpha_{i0};\theta_0}$ is a density with respect to an appropriate dominating measure (Lebesgue, counting, or a mixture of both), which does not depend on $(_{iT}, \alpha_{i0}, \theta_0)$. The dominating measure is not explicitly specified.

Assumption 6.2 (*Identification*) θ_0 is identified, i.e., uniquely defined. □

In particular, θ_0 is identified as the well-separated global maximum of the limit (as $n, T \to \infty$) of the expected "target" loglikelihood (defined in Sect. 6.4.2) for the sample. Assumption 6.2 rules out time-invariant explanatory variables in $_{iT}$ that have a time-invariant relationship with α_{i0}.

Assumption 6.3 (*Sampling*) (i) For each T, $(_{1T}, _{1T}, \alpha_{10}), \ldots, (_{nT}, _{nT}, \alpha_{n0})$ are independently and identically distributed (i.i.d.); (ii) For each i, conditional on $_{iT}, \alpha_{i0}$, the outcomes $Y_{i1}, \ldots, Y_{iT}$ are independent; (iii) For each i, conditional on α_{i0}, the process $(Y_{it}, X_{it})_{t \in \mathbb{N}}$ is strictly stationary. □

Part (i) of Assumption 6.3, which stipulates that observations across i are i.i.d., is typical in microeconometric applications and is maintained throughout the chapter. Part (ii) imposes conditional independence within each i, hereafter referred to as "time-independence," which rules out lagged outcomes as explanatory variables. However, time-independence is not necessary for the methodology developed in this chapter to work. To illustrate this, we apply our approach to the linear dynamic panel data model in Sect. 6.12 without imposing time-independence. Part (iii) rules out the presence of time-varying parameters or time-trends in the model for the outcomes, i.e., there are no time-varying parameters or time-trends in $f_{ _{iT}| _{iT},\alpha_{i0};\theta_0}$. Allowing for time-varying parameters in our approach is, in principle, possible. However, doing so will significantly increase the technical complexity of the proofs in the $n, T \to \infty$ setting, cf., e.g., Fernández-Val and Weidner (2016).

Since α_{i0} is an unobserved random variable, we may speak of the likelihood corresponding to any potential realization. Specifically, if $\theta \in \Theta$ and $\alpha_i \in (\mathfrak{a}, \mathfrak{b})$ denotes a possible value taken by α_{i0}, then we define the likelihood of (θ, α_i) for the ith individual to be $L_{iT}(\theta, \alpha_i) := f_{\ \ iT|\ \ iT,\alpha_i;\theta}(\ \ iT)$. The average loglikelihood for the ith individual is denoted by $\ell_{iT}(\theta, \alpha_i) := T^{-1} \log L_{iT}(\theta, \alpha_i)$. We refer to θ as the parameter of interest, and call α_i an individual-specific nuisance parameter. The loglikelihood function $(\theta, \alpha_i) \mapsto \ell_{iT}(\theta, \alpha_i)$ is assumed to be sufficiently regular so that derivatives with respect to (θ, α_i), as many as needed, can be interchanged with integrals respect to the density $f_{\ \ iT|\ \ iT,\alpha_i;\theta}$, and that mixed partial derivatives are equal.

The score of $\ell_{iT}(\theta, \alpha_i)$ with respect to θ is the (column) vector $\ell_{iT\theta}(\theta, \alpha_i) := \nabla_\theta \ell_{iT}(\theta, \alpha_i)$, where $\nabla_\theta := (\partial_\theta)'$ is the gradient and "$'$" the transpose operator. As α_i is a scalar, $\ell_{iT\alpha}(\theta, \alpha_i) := \nabla_\alpha \ell_{iT}(\theta, \alpha_i) = \partial_\alpha \ell_{iT}(\theta, \alpha_i)$ denotes the score with respect to α_i. We use $f_{ab} := \partial_b \circ \nabla_a f$ to denote mixed partial derivatives of second order. Consequently, $\ell_{iT\theta\theta}(\theta, \alpha_i)$ is a square matrix, $\ell_{iT\theta\alpha}(\theta, \alpha_i)$ is a column vector, $\ell_{iT\alpha\theta}(\theta, \alpha_i)$ is a row vector (with $\ell'_{iT\alpha\theta} = \ell_{iT\theta\alpha}$), and $\ell_{iT\alpha\alpha}(\theta, \alpha_i)$ is a scalar.

Given $(\check{\theta}, \check{\alpha}) \in \Theta \times (\mathfrak{a}, \mathfrak{b})$, let $\mathbb{E}[\ell_{iT\alpha}(\theta, \alpha_i); \check{\theta}, \check{\alpha}] := \int_{\text{supp}(\ \ iT)} \ell_{iT\alpha}(\theta, \alpha_i)\, f_{\ \ iT|\ \ iT,\check{\alpha};\check{\theta}}$, where supp($_{iT}$) denotes the support of $\ \ _{iT}$, and integration is with respect to the (unspecified) dominating measure for f. The integral can be calculated analytically or numerically, depending on the functional form of f. Integration with respect to $f_{\ \ iT|\ \ iT,\alpha_{i0};\theta_0}$ is denoted by $\mathbb{E}_0$, e.g., $\mathbb{E}_0 \ell_{iT\alpha}(\theta, \alpha_i) := \mathbb{E}[\ell_{iT\alpha}(\theta, \alpha_i); \theta_0, \alpha_{i0}]$. Similarly, $\text{var}_0\, \ell_{iT\alpha}(\theta, \alpha_i) := \mathbb{E}_0[\ell_{iT\alpha}(\theta, \alpha_i) - \mathbb{E}_0 \ell_{iT\alpha}(\theta, \alpha_i)]^2$. We usually omit the arguments in functionals of $\mathbb{E}_0 \ell_{iT}(\theta, \alpha_i)$ when evaluated at (θ_0, α_{i0}). E.g., we write $\mathbb{E}_0 \ell^2_{iT\alpha} := \mathbb{E}_0 \ell^2_{iT\alpha}(\theta_0, \alpha_{i0})$, $\mathbb{E}_0 \ell_{iT\alpha\alpha} := \mathbb{E}_0 \ell_{iT\alpha\alpha}(\theta_0, \alpha_{i0})$, etc. Since $\mathbb{E}_0$ is a conditional expectation and var_0 a conditional variance (both conditional on $\ \ _{iT}, \alpha_{i0}$), equalities and inequalities involving them hold w.p.1. To avoid the proliferation of "w.p.1" qualifiers each time $\mathbb{E}_0$ or var_0 is mentioned, we do not state them explicitly hereafter.

Henceforth, to allow the number of time periods to grow simultaneously with the number of individuals, let (T_n) be a sequence of positive integers such that $T_n \to \infty$ as $n \to \infty$. When there is no danger of confusion, we do not indicate the dependence of estimators on n and T. If A_{iT} is an array, then the statement $A_{iT} = O_{\text{p}}(1)$ is understood to hold coordinatewise.

6.4 Problem with the Maximum Likelihood Estimator (MLE)

Since the distribution of $\ \ _{iT}|\ \ _{iT}, \alpha_i$ is known up to (θ, α_i), it is natural to estimate θ by maximum likelihood while treating the α_i as nuisance parameters. The fixed-effects MLE of θ is given by

$$\tilde{\theta} := \operatorname*{argmax}_{\theta \in \Theta} \max_{\alpha_1, \ldots, \alpha_n \in (\mathfrak{a}, \mathfrak{b})} n^{-1} \sum_{i=1}^{n} \ell_{iT}(\theta, \alpha_i). \tag{6.1}$$

Unfortunately, since the number of nuisance parameters grows with the number of individuals, the MLE of θ may not be consistent when $n \to \infty$ and T is fixed. The following example, due to Neyman and Scott (1948), illustrates this phenomenon beautifully.

6.4.1 Heterogeneous Means Model of Neyman and Scott

Consider a Gaussian panel data model where individual-specific heterogeneity arises only in the mean, i.e., let $Y_{it} = \alpha_{i0} + U_{it}$, where $U_{i1}, \ldots, U_{iT} \mid \alpha_{i0} \stackrel{\text{d}}{=} \text{NIID}(0, \sigma_0^2)$ with $\text{supp}(\alpha_{i0}) = \mathbb{R}$ and $\sigma_0^2 \in (0, \infty) =: \Theta$. The Neyman-Scott model has been used to explain variation in scores $(Y_{i1}, \ldots, Y_{iT})$ obtained by individual i on repeated IQ tests using unobserved individual ability (α_{i0}) as the only explanatory variable. The parameter of interest is the variance of the IQ scores, i.e., $\theta_0 := \sigma_0^2$. The fixed-effects MLE of σ_0^2 is given by the average (across individuals) of the within-individual sample variances, i.e.,

$$\tilde{\sigma}^2 := \frac{1}{n}\sum_{i=1}^{n}\frac{1}{T}\sum_{t=1}^{T}(Y_{it} - \bar{Y}_{i\cdot})^2 \stackrel{\text{p}}{\to} \sigma_0^2\frac{(T-1)}{T} \quad \text{as } n \to \infty \text{ and } T \text{ is held fixed.}$$

Therefore, $\tilde{\sigma}^2$ is not consistent for σ_0^2 as $n \to \infty$ and T is fixed. This is the classic "incidental parameters problem' of the fixed-effects MLE, first noted by Neyman and Scott. SST21 show that

$$\tilde{\sigma}^2 - \sigma_0^2 = O_{\text{p}}\left(\frac{1}{\sqrt{nT}}\right) + O_{\text{p}}\left(\frac{1}{T}\right) \quad \text{as } n, T \to \infty.$$

Hence, the fixed-effects MLE is consistent as $n, T \to \infty$, irrespective of whether T grows faster or slower than n. Furthermore, as $n, T_n \to \infty$,

$$\sqrt{nT_n}(\tilde{\sigma}^2 - \sigma_0^2) \stackrel{\text{d}}{\to} \begin{cases} \text{N}(-\sigma_0^2\sqrt{\rho}, 2\sigma_0^4) & \text{if } \rho := \lim_{n\to\infty} n/T_n \in (0, \infty) \\ \text{N}(0, 2\sigma_0^4) & \text{if } \lim_{n\to\infty} n/T_n = 0 \\ -\infty & \text{if } \lim_{n\to\infty} n/T_n = \infty. \end{cases} \tag{6.2}$$

Therefore, as noted by Li, Lindsay, and Waterman (2003, Sect. 2), even though $\tilde{\sigma}$ is consistent, the asymptotic distribution of $\sqrt{nT_n}(\tilde{\sigma}^2 - \sigma_0^2)$ is not centered at the origin if n and T_n grow at the same rate.[1] More generally, Li et al. and Hahn and Newey (2004) have shown that if n and T_n grow at the same rate, then the fixed-effects MLE defined in (6.1) is asymptotically biased in the sense that $\sqrt{nT_n}(\tilde{\theta} - \theta_0)$ converges in distribution to a Gaussian random vector whose mean is not zero.

6.4.2 Source of the Incidental Parameters Problem

Since each individual contributes a single individual-specific nuisance parameter, the MLE in (6.1) can be written as $\tilde{\theta} = \text{argmax}_{\theta\in\Theta} n^{-1} \sum_{i=1}^{n} \ell_{iT}^{\text{p}}(\theta)$, where $\ell_{iT}^{\text{p}}(\theta) := \ell_{iT}(\theta, \hat{\alpha}_{iT}(\theta))$ is the average profile loglikelihood of θ for individual i after the nuisance parameters have been profiled out, and

$$\hat{\alpha}_{iT}(\theta) := \underset{u\in(\mathfrak{a},\mathfrak{b})}{\text{argmax}}\, \ell_{iT}(\theta, u), \qquad \theta \in \Theta,$$

is the MLE of α_i for a given θ. For future reference, we also let

$$\alpha_{iT}^{*}(\theta) := \underset{u\in(\mathfrak{a},\mathfrak{b})}{\text{argmax}}\, \mathbb{E}_0 \ell_{iT}(\theta, u), \qquad \theta \in \Theta,$$

and refer to it as the "population level MLE" of α_i for a given θ. Following Pace and Salvan (2006, Sect. 3.2), we refer to $\ell_{iT}(\theta, \alpha_{iT}^{*}(\theta))$ as the "target" loglikelihood of θ for the ith individual. Note that the (infeasible) target loglikelihood is a genuine loglikelihood function; in particular, it satisfies all of the Bartlett identities.

The inconsistency of $\tilde{\theta}$ (when $n \to \infty$, T is fixed, and the profile-likelihood is not a conditional likelihood free of nuisance parameters, as in the panel Poisson example in Sect. 6.12.2), is due to the fact that the individual-specific nuisance parameters are poorly estimated when T is held fixed. Indeed, from its definition it is clear that $\hat{\alpha}_{iT}(\theta_0)$ does not depend on n. Therefore, when $\hat{\alpha}_{iT}(\theta_0) \neq \alpha_{i0}$, $\hat{\alpha}_{iT}(\theta_0)$ will not converge to α_{i0} when $n \to \infty$ and T is fixed. This is the sense in which the individual-specific parameters are poorly estimated when $n \to \infty$ but T is fixed. This implies that the profile-likelihood scores for each individual do not have zero mean when evaluated at the true parameter values (McCullagh and Tibshirani 1990, Remark 2, p. 329). In fact, SST21 show that

$$\mathbb{E}_0 \nabla_\theta \ell_{iT}^{\text{p}}(\theta_0) = O_{\text{p}}(T^{-1}), \tag{6.3}$$

and that $\tilde{\theta}$ is inconsistent, as $n \to \infty$ and T-fixed, where $\nabla_\theta \ell_{iT}^{\text{p}}(\theta) = \ell_{iT\theta}(\theta, \hat{\alpha}_{iT}(\theta))$ is the profile-likelihood score of θ for the ith individual.

---

[1] Although the asymptotic bias disappears if T_n grows faster than n, this asymptotic setting is not relevant for modeling microeconometric panels, which, as mentioned earlier, typically consist of a large number of individuals followed for a comparatively smaller number of time periods.

Since the score function of a genuine loglikelihood has zero mean, (6.3) reveals that the bias of $\nabla_\theta \ell_{iT}^{\mathrm{p}}(\theta_0)$ is of the order $1/T$ for each i. Hence, allowing T to grow (along with n) may enable $\tilde{\theta}$ to consistently estimate its true value θ_0 as both $n, T \to \infty$. However, as is clear from the Neyman-Scott model in Sect. 6.4.1, this alone may not be sufficient to ensure that $\tilde{\theta}$ is asymptotically unbiased in the sense that the limiting distribution of $\sqrt{nT_n}(\tilde{\theta} - \theta_0)$ is correctly centered at the origin.

6.5 Addressing the Problem of the Fixed-Effects MLE

Efforts to resolve this issue, by developing estimators with correctly centered asymptotic distributions and asymptotic variances identical to that of the fixed-effects MLE, have produced an extensive literature. E.g., Lancaster (2002) has suggested an IL approach based on orthogonalizing the parameter of interest and the individual-specific nuisance parameter using the "information orthogonalizing transformation (IOT)"; cf. Cox and Reid (1987) and Severini (2000, Sect. 3.6.4) on how the IOT is obtained. Lancaster's approach can be restrictive because finding an IOT requires solving a differential equation; when θ is a scalar, a solution always exists, though finding it may be challenging. When θ is a vector, such a solution may not exist, so that a IOT may not exist. Important applications in which this situation arises include the AR(1) model with covariates (Lancaster 2002, Sect. 3.2) and autoregressive models of order greater than one. The latter case was conjectured by Lancaster (2002, p. 663), who did not provide a proof; a formal proof is given by Dhaene and Jochmans (2016, p. 1208).

It is thus desirable to obtain estimators of θ that do not require the IOT, so that they are applicable in general situations where θ is a vector. One approach is to employ the jackknife or analytical bias corrections. Cf., e.g., Hahn and Kuersteiner (2002), Woutersen (2002), Arellano (2003a), Hahn and Newey (2004), Arellano and Hahn (2007), Carro (2007), Bester and Hansen (2009), Fernández-Val (2009), Hahn and Kuersteiner (2011), and Dhaene and Jochmans (2015). Alternatively, Arellano and Bonhomme (2009), henceforth AB, propose an IL approach that does not require θ and α_i to be orthogonal. Their estimator is $\hat{\theta}_{\mathrm{AB}} := \operatorname{argmax}_{\theta \in \Theta} n^{-1} \sum_{i=1}^{n} \bar{\ell}_{iT}^{\mathrm{AB}}(\theta)$, where $\bar{\ell}_{iT}^{\mathrm{AB}}(\theta) := T^{-1} \log \int L_{iT}(\theta, \alpha) \hat{w}_i(\theta, \alpha)\, d\alpha$ is the log-IL of AB, and $\hat{w}_i$ is an individual-specific data-dependent weight-function. The weights $\hat{w}_1, \ldots, \hat{w}_n$ are chosen such that the bias of the IL score for each i, when each $\hat{w}_i$ is replaced by its population counterparts and all parameters are evaluated at the truth, is of the order $1/T^2$ as $T \to \infty$. For each i, the population scores in AB's approach can therefore be regarded as being "first-order unbiased" as compared to the profile-likelihood scores, whose bias is only of the order $1/T$ (cf. (6.3)). Under the condition that $\lim_{n\to\infty} n/T_n \in (0, \infty)$, $\sqrt{nT_n}(\hat{\theta}_{\mathrm{AB}} - \theta_0)$ is asymptotically normal with mean zero and variance equal to that of the fixed-effects MLE. Recent works similar to AB include De Bin, Sartori, and Severini (2015) and Pakel (2019), with the latter allowing for both time series and cross-sectional dependence.

SST21 construct an IL that behaves like the target likelihood for estimating the parameter of interest. Their IL is based on extending the approach in Severini (2007) to panel data models. Specifically, the maximum integrated likelihood estimator (MILE) they propose is based on a certain data-dependent transformation of α_i, called the "zero-score-expectation (ZSE)" transformation, which is used to construct an IL possessing desirable properties, irrespective of the weight-function used to integrate out the transformed nuisance parameter. The ZSE transformation ensures that, regardless of the weight-function, their IL is closer to the target likelihood, i.e., the Bartlett identities for it are closer to being satisfied, which has positive implications for estimation and inference.

The usual approach to determine whether a random function behaves like a genuine likelihood is to check if it satisfies the Bartlett identities, particularly score unbiasedness (the 1st Bartlett identity) and information unbiasedness (the 2nd Bartlett identity). In standard parametric panel data models, the profile-likelihood satisfies both identities with error $O(1/T)$. The IL of AB and Lancaster satisfy the first identity with error $O(1/T^2)$, whereas the second identity holds only up to an error of order $O(1/T)$; cf. Sect. 6.9.4. In contrast, the ZSE transformed IL satisfies both identities with error $O(1/T^2)$; cf. Sects. 6.9.3 and 6.9.4. That is, unlike the ILs of AB and Lancaster, the IL of SST21 is simultaneously first-order score and first-order information unbiased. There is therefore a clear and intuitive sense in which the ZSE transformed IL improves upon the earlier approaches.

6.6 The ZSE Transformation

The fixed-effects MLE of θ is inconsistent because estimation of α_i influences the estimation of θ. We fix this problem by transforming α_i—using the ZSE transformation of Severini (2007) defined subsequently—into a new "functional" nuisance parameter that is "strongly unrelated" to θ (cf. Definition 6.1).[2] This leads to a transformed likelihood, from which the new nuisance parameter is eliminated by integrating it out. This IL is then used to construct an estimator of θ having the desired properties. As mentioned earlier, unlike AB, the choice of weight-functions used to integrate out the nuisance parameter is not critical in our approach.

Definition 6.1 (*Strong unrelatedness*) In the loglikelihood $\ell_{iT}(\theta, \alpha_i)$, the individual-specific nuisance parameter α_i is said to be strongly unrelated at the population level to the parameter of interest θ if $\alpha^*_{iT}(\theta) = \alpha_{i0}$ for each $\theta \in \Theta$. □

In other words, α_i is strongly unrelated to θ at the population level if its population level MLE (for fixed θ) does not depend on θ. Severini (2007, p. 530) defines the strong unrelatedness property in terms of estimators. Definition 6.1 is the analogous version in terms of population level parameters.

[2] Since α_i is individual-specific, we are free to transform it provided the transformed value is also individual-specific so that the interpretation of the model is not altered. Strong unrelatedness of the parameters has several important consequences. Cf. the discussion after Lemma 6.1, and Sect. 6.9, for more on this.

We begin by giving some intuition behind the ZSE transformation; cf. Severini (2007) for further discussion. Consider a statistical model with parameter of interest θ; suppose that it is possible to find a nuisance parameter λ such that the likelihood function $L(\theta, \lambda)$ can be factored as

$$L(\theta, \lambda) = L_1(\theta)L_2(\lambda), \tag{6.4}$$

for functions L_1, L_2. Note that, when the factorization (6.4) holds, the parameters θ, λ are unrelated in a certain sense; for instance, the MLE of λ for fixed θ is simply the overall MLE of λ.

When (6.4) holds, inference for θ can be based on $L_1(\theta)$. Furthermore, because the corresponding loglikelihood for (θ, λ) satisfies the Bartlett identities, and it can be written as a function of θ plus a function of λ, $\log L_1(\theta)$ satisfies the Bartlett identities. Note that $L_1(\theta)$ can be obtained from $L(\theta, \lambda)$ as an integrated likelihood function using any weight-function for λ that does not depend on θ. This suggests that, for inference for a parameter of interest in the presence of a nuisance parameter, if a nuisance parameter can be found that is approximately unrelated, in some sense, to the parameter of interest, then the integrated likelihood based on a weight-function for the nuisance parameter that does not depend on the parameter of interest should satisfy the Bartlett identities to a high degree of approximation. Furthermore, any weight-function for the nuisance parameter that does not depend on the parameter of interest yields approximately the same integrated likelihood function (up to a multiplicative constant).

These considerations motivate the ZSE transformation, which, loosely speaking, is a data-based bijective function $\alpha_i \mapsto g(\alpha_i)$ such that $g(\alpha^*_{iT}(\theta))$, the transformed population level MLE of α_i, does not depend on θ. How can such a mapping be constructed? Observe that the first-order condition (FOC) for $\alpha^*_{iT}(\theta)$ is $\mathbb{E}[\ell_{iT\alpha}(\theta, \alpha^*_{iT}(\theta)); \theta_0, \alpha_{i0}] = 0$. Hence, if g is defined as solving $\mathbb{E}[\ell_{iT\alpha}(\theta, \alpha_i); \theta_0, g(\alpha_i))] = 0$ for each α_i, then $g(\alpha^*_{iT}(\theta)) = \alpha_{i0}$, i.e., $g(\alpha^*_{iT}(\theta))$ does not depend on θ as desired. Although g constructed in this manner depends generally on θ_0 (which is unknown), a feasible version of g can be obtained by replacing θ_0 by a preliminary estimator, e.g., the fixed-effects MLE, which is consistent as $n, T \to \infty$. We now make these notions precise.

Let $g_{iT\theta_0\theta} : (\mathfrak{a}, \mathfrak{b}) \to (\mathfrak{a}, \mathfrak{b})$ denote a function, which depends on i, T, θ_0, θ, such that for all $\alpha_i \in (\mathfrak{a}, \mathfrak{b})$, $g_{iT\theta_0\theta}(\alpha_i)$ satisfies $\mathbb{E}[\ell_{iT\alpha}(\theta, \alpha_i); \theta_0, g_{iT\theta_0\theta}(\alpha_i)] = 0$. Following Severini (2007, p. 532), we refer to $\alpha \mapsto g_{iT\theta_0\theta}(\alpha)$ as the ZSE transformation and $\phi := g_{iT\theta_0\theta}(\alpha)$ as the ZSE (nuisance) parameter corresponding to α for a given θ. The local existence and uniqueness of the mapping $\alpha \mapsto g_{iT\theta_0\theta}(\alpha)$ and its inverse $\phi \mapsto h_{iT\theta_0\theta}(\phi)$ — for θ in a neighborhood of θ_0 — which is required to define the transformed likelihood in terms of the ZSE parameter, follows from Lemma 4.1 in SST21.

If the optimization problem $\max_u \mathbb{E}[\ell_{iT}(\theta, u); \theta_0, \phi]$ has a unique solution then that solution must be $h_{iT\theta_0\theta}(\phi)$. In other words, $h_{iT\theta_0\theta}(\phi)$ is the population level MLE of α_i when the true value of (θ, α_i) is (θ_0, ϕ). Extending this analogy, if the population level MLE of α_i exists for all true values of (θ, α_i), i.e., if the optimization problem $\max_u \mathbb{E}[\ell_{iT}(\theta, u); \theta_0, \phi]$ has a unique solution for each (θ, θ_0, ϕ)—which, e.g., is the case if $u \mapsto \mathbb{E}[\ell_{iT}(\theta, u); \theta_0, \phi]$ is strictly concave for each (θ, θ_0, ϕ)—then, for each θ_0, θ, the inverse ZSE transformation $h_{iT\theta_0\theta}$ exists. This helps explain why the inverse ZSE transformation can exist even when the IOT does not.[3]

Interpreting $h_{iT\theta_0\theta}$ as a population level MLE justifies its global existence, i.e., for all $\theta_0, \theta \in \Theta$. The inverse ZSE transformation is then used to construct $\tilde{L}^0_{iT}(\theta, \phi) := L_{iT}(\theta, h_{iT\theta_0\theta}(\phi))$, the infeasible ZSE transformed likelihood for individual i, where infeasibility arises from its dependence on θ_0. The corresponding loglikelihood, $\tilde{\ell}^0_{iT}(\theta, \phi) := T^{-1} \log \tilde{L}^0_{iT}(\theta, \phi) = \ell_{iT}(\theta, h_{iT\theta_0\theta}(\phi))$, is useful because of the strong unrelatedness property of $\phi^*_{iT}(\theta)$, the population level MLE of the ZSE parameter, defined by

$$\phi^*_{iT}(\theta) := \underset{\phi \in (a,b)}{\operatorname{argmax}}\, \mathbb{E}_0 \tilde{\ell}^0_{iT}(\theta, \phi), \qquad \theta \in \Theta.$$

As established in Lemma 6.1, this strong unrelatedness property is precisely what makes the ZSE transformation so effective.

Lemma 6.1 *(Strong unrelatedness property of the ZSE parameter; SST21, Lemma 4.2)* $\phi^*_{iT}(\theta) = \alpha_{i0}$ *for each* $\theta \in \Theta$. □

Lemma 6.1 reveals that, in the infeasible ZSE transformed loglikelihood $\tilde{\ell}^0_{iT}(\theta, \phi)$, the ZSE parameter ϕ is strongly unrelated to the parameter of interest θ at the population level. In asymptotic expansions, strong unrelatedness of ϕ and θ allows $\phi^*_{iT}(\theta)$ to be replaced by α_{i0} without creating bias. The strong unrelatedness property also suggests that eliminating ϕ from $\tilde{\ell}^0_{iT}(\theta, \phi)$ will not affect the estimation of θ. Indeed, it is the strong unrelatedness of ϕ and θ that reduces both score and information bias for the ZSE transformed IL.

It is worth noting that a nuisance parameter ϕ that is information orthogonal to θ satisfies a local version of the result of Lemma 6.1: $\phi^*_{iT}(\theta) = \alpha_{i0} + O_p(\|\theta - \theta_0\|^2)$ (Severini 2000, Sect. 4.6). Thus, it is not surprising that a consequence of ϕ and θ being strongly unrelated is that they are information orthogonal, i.e., $\mathbb{E}_0 \nabla^2_{\theta\phi} \tilde{\ell}^0_{iT}(\theta_0, \alpha_{i0}) = 0$, where $\nabla^2_{ab} := \partial_b \circ \nabla_a$ and $\nabla^2_{\theta\phi} \tilde{\ell}^0_{iT}(\theta_0, \alpha_{i0}) := \nabla^2_{\theta\phi} \tilde{\ell}^0_{iT}(\theta, \phi)\big|_{\theta=\theta_0, \phi=\alpha_{i0}}$. This shows that the ZSE transformation makes θ and ϕ information orthogonal,[4] even though it is not the IOT, as it is characterized differently. Moreover, the example in Sect. 6.6.1 shows that a nuisance parameter may

---

[3] The differential equations that define the IOT may not be solvable if $\dim(\theta) > 1$ (Cox and Reid, Sect. 2.3). Hence, the IOT is not guaranteed to exist if θ is a vector. However, as the inverse ZSE transformation is characterized differently, it can exist even when the IOT does not. For instance, although the IOT does not exist for an AR(1) model with covariates (Lancaster 2002, Sect. 3.2), the calculations for the dynamic Neyman-Scott model in Sect. 6.12.5 can be extended to show that $h_{iT\theta_0\theta}$ does exist in this model.

[4] The ZSE transformation is not merely another device for orthogonalizing parameters. It accomplishes considerably more: rather than simply rendering the parameters orthogonal in the transformed likelihood, it makes them strongly unrelated. It is strong unrelatedness—rather than information orthogonality alone—that drives the reduction in both score and information bias for the ZSE transformed IL.

be information orthogonal to the parameter of interest without being strongly unrelated to it, implying that the ZSE transformation and the IOT are fundamentally distinct.

6.6.1 A Panel Model with Heterogeneous Variances

Consider a simple Gaussian panel data model where the heterogeneity arises not in the mean (as in the Neyman-Scott model), but in the variance. That is, let $Y_{it} = \theta_0 + U_{it}$, where $U_{i1}, \ldots, U_{iT} | \sigma_{i0}^2 \overset{\mathrm{d}}{=} \text{NIID}(0, \sigma_{i0}^2)$, so that the individual-specific nuisance parameter is now $\alpha_i := \sigma_i^2$. In this example, α_i and θ are information orthogonal. However, α_i is not strongly unrelated to θ at the population level because it is straightforward to verify that $\alpha_{iT}^*(\theta)$ is not a constant function of θ. Indeed, $\alpha_{iT}^*(\theta) = \sigma_{i0}^2 + (\theta - \theta_0)^2 \neq \sigma_{i0}^2$ in each deleted neighborhood of θ_0. It is interesting to note that in the Neyman-Scott model the individual-specific nuisance parameter is information orthogonal, as well as strongly unrelated at the population level, to the parameter of interest.

6.7 The MILE

Although the inverse ZSE transformation can be determined analytically in certain cases, cf. the static Neyman-Scott, panel Poisson, and dynamic Neyman-Scott examples in Sects. 6.12.1, 6.12.2, and 6.12.5, it is typically obtained numerically as in panel Logit and panel Probit examples in Sects. 6.12.3 and 6.12.4. In principle, this is straightforward to do by fixing θ and then, for each given ϕ, finding a number h that numerically solves the equation $\mathbb{E}[\ell_{iT\alpha}(\theta, h); \theta_0, \phi] = 0$. In practice, however, θ_0 is first replaced by a preliminary estimator, e.g., the fixed-effects MLE $\tilde{\theta}$, which is consistent as $n, T \to \infty$. Then, given ϕ, the equation $\mathbb{E}[\ell_{iT\alpha}(\theta, h); \tilde{\theta}, \phi] = 0$ is solved numerically for h. The solution is the function $\phi \mapsto h_{iT\tilde{\theta}\theta}(\phi)$, the estimator of the inverse ZSE transformation $\phi \mapsto h_{iT\theta_0\theta}(\phi)$.

Let $\tilde{L}_{iT}(\theta, \phi) := L_{iT}(\theta, h_{iT\tilde{\theta}\theta}(\phi))$ denote the feasible version of $\tilde{L}_{iT}^0(\theta, \phi)$. Similarly, $\tilde{\ell}_{iT}(\theta, \phi) := T^{-1} \log \tilde{L}_{iT}(\theta, \phi) = \ell_{iT}(\theta, h_{iT\tilde{\theta}\theta}(\phi))$ is the feasible version of $\tilde{\ell}_{iT}^0(\theta, \phi)$. The feasible ZSE transformed IL for $\theta \in \Theta$ for the ith individual is defined to be

$$\bar{L}_{iT}(\theta) := \int_{(\mathfrak{a}, \mathfrak{b})} \tilde{L}_{iT}(\theta, \phi)\pi_i(\phi)\, d\phi = \int_{(\mathfrak{a}, \mathfrak{b})} L_{iT}(\theta, h_{iT\tilde{\theta}\theta}(\phi))\pi_i(\phi)\, d\phi, \tag{6.5}$$

where $\pi_i : (\mathfrak{a}, \mathfrak{b}) \to (0, \infty)$ is a weight-function that does not depend on θ, and it is assumed that the integral in (6.5) is finite for each $\theta \in \Theta$. Unlike AB, the choice of π_i here is not critical. Indeed, since $\nabla_\theta \bar{L}_{iT}(\theta)$ can be shown to be approximately independent of π_i (Sect. 6.9.2), it is acceptable to let $\pi_i := 1$, which is what we do in the examples in Sect. 6.12.

Let $\bar{\ell}_{iT}(\theta) := T^{-1} \log \bar{L}_{iT}(\theta)$ denote the ZSE transformed log-IL for individual i. The MILE of θ is defined to be $\hat{\theta} := \operatorname{argmax}_{\theta\in\Theta} n^{-1} \sum_{i=1}^{n} \bar{\ell}_{iT}(\theta)$, i.e.,[5]

$$\hat{\theta} := \operatorname*{argmax}_{\theta\in\Theta} n^{-1} \sum_{i=1}^{n} T^{-1} \log \int_{(\mathfrak{a},\mathfrak{b})} L_{iT}(\theta, h_{iT\tilde{\theta}\theta}(\phi))\pi_i(\phi)\, d\phi. \tag{6.6}$$

The definition of the MILE makes clear the difference between $\hat{\theta}$ and $\hat{\theta}_{\mathrm{AB}}$. Namely, we use the data-dependent ZSE transformation of the nuisance parameter to define our IL, whereas AB find a data-dependent weight-function for the nuisance parameter to define their IL.

The MILE can be iterated to remove its dependence on the preliminary estimator $\tilde{\theta}$, and perhaps even improve its finite-sample properties: Once $\hat{\theta}$ becomes available, it is used to obtain $h_{iT\hat{\theta}\theta}$, which is then employed to recompute the MILE as defined in (6.6). This yields the single-iteration MILE $\hat{\theta}(1) := \operatorname{argmax}_{\theta\in\Theta} n^{-1} \sum_{i=1}^{n} T^{-1} \log \int_{(\mathfrak{a},\mathfrak{b})} L_{iT}(\theta, h_{iT\hat{\theta}\theta}(\phi))\pi_i(\phi)\, d\phi$. And this process can be repeated until convergence.

Remark 6.1 (*Maximization vs. integration*) As with any interest respecting transformation (Sect. 6.9.1), maximizing the ZSE transformed loglikelihood $\tilde{\ell}_{iT}(\theta, \phi)$ with respect to ϕ yields the same profile loglikelihood as in the original parameterization. Specifically, the profile loglikelihood of $\tilde{\ell}_{iT}(\theta, \phi)$, given by $\tilde{\ell}_{iT}(\theta, \hat{\phi}_{iT\tilde{\theta}}(\theta))$, where

$$\hat{\phi}_{iT\tilde{\theta}}(\theta) := \operatorname*{argmax}_{\phi\in(\mathfrak{a},\mathfrak{b})} \ell_{iT}(\theta, h_{iT\tilde{\theta}\theta}(\phi)), \qquad \theta \in \Theta,$$

coincides with the original profile loglikelihood, i.e.,

$$\tilde{\ell}_{iT}(\theta, \hat{\phi}_{iT\tilde{\theta}}(\theta)) = \ell_{iT}(\theta, \hat{\alpha}_{iT}(\theta)), \qquad \theta \in \Theta.$$

Consequently, in the definition of the MILE, it is essential to integrate out the ZSE parameter rather than maximize over it. Indeed, the optimization problem $\operatorname{argmax}_{\theta\in\Theta} \sum_{i=1}^{n} \max_{\phi\in(\mathfrak{a},\mathfrak{b})} \tilde{\ell}_{iT}(\theta, \phi)$ simply yields the fixed-effects MLE. □

[5] Although the ZSE transformed IL is constructed using the fixed-effects MLE, whose asymptotic distribution is biased as $n, T \to \infty$, the limiting distribution of the MILE is correctly centered as $n, T \to \infty$. Consequently, the MILE is not hindered by the fact that the preliminary estimator used to construct $h_{iT\tilde{\theta}\theta}$, namely, the fixed-effects MLE, is asymptotically biased.

6.8 Approximating the ZSE Transformed IL

To get some insight behind the structure of the ZSE transformed IL we can use the following result, which provides a uniform (in θ) Laplace approximation of $\bar{\ell}_{iT}(\theta)$. Throughout this section, if not stated explicitly, limits are taken as $T \to \infty$.

Lemma 6.2 *(Laplace approximation of the ZSE transformed log-IL; SST21, Lemma 6.1) Let* $c_T := (2T)^{-1}\log(2\pi/T)$ *and* $\tilde{\ell}_{iT\phi\phi}(\theta,\phi) := \partial^2_\phi \tilde{\ell}_{iT}(\theta,\phi)$. *Then,*

$$\bar{\ell}_{iT}(\theta) = c_T + \ell_{iT}(\theta, \hat{\alpha}_{iT}(\theta)) - \frac{1}{2T}\log(-\tilde{\ell}_{iT\phi\phi}(\theta, \hat{\phi}_{iT\tilde{\theta}}(\theta))) + \frac{1}{T}\log \pi_i(\hat{\phi}_{iT\tilde{\theta}}(\theta)) + R_{iT}(\theta),$$

where $\sup_{\theta\in\Theta}|R_{iT}(\theta)| = O_p(T^{-2})$ *as* $T \to \infty$. □

It is clear from Lemma 6.2 that $\bar{\ell}_{iT}(\theta)$, modulo a constant and an $O_p(T^{-2})$ remainder term, is the sum of the profile loglikelihood, a term $\log(-\tilde{\ell}_{iT\phi\phi}(\theta, \hat{\phi}_{iT\tilde{\theta}}(\theta)))$ that reflects how the ZSE transformation "additively corrects" the profile loglikelihood, and the weight-function $\log \pi_i(\hat{\phi}_{iT\tilde{\theta}}(\theta))$. It is the correction term $\log(-\tilde{\ell}_{iT\phi\phi}(\theta, \hat{\phi}_{iT\tilde{\theta}}(\theta)))$, which includes a contribution from the inverse ZSE transformation, that causes the IL, hence, the MILE, to possess the desired properties. In contrast, the weight-function has no real effect on the MILE.

The correction term $\log(-\tilde{\ell}_{iT\phi\phi}(\theta, \hat{\phi}_{iT\tilde{\theta}}(\theta)))$ can be contrasted with the adjustment to the profile loglikelihood in Cox and Reid (1987, Equation 10). Whereas Cox and Reid use the IOT to transform the nuisance parameter, SST21 use the ZSE transformation, which has the advantage, relative to the IOT, that it can exist even when the IOT does not, and that it is invariant to interest respecting reparametrizations (Sect. 6.9.1).

Lemma 6.2 has a useful corollary that can provide additional insights. Cf. the panel logit example in Sect. 6.12.3 for an illustration.

Corollary 6.1 *(Laplace approximation of the ZSE transformed IL; SST21, Corollary 6.1) Under the assumptions of Lemma 6.2,*

$$\bar{L}_{iT}(\theta) = \sqrt{\frac{2\pi}{T}}\frac{L_{iT}(\theta, \hat{\alpha}_{iT}(\theta))}{\sqrt{-\ell_{iT\alpha\alpha}(\theta, \hat{\alpha}_{iT}(\theta))}}\frac{\pi_i(\hat{\phi}_{iT\tilde{\theta}}(\theta))}{|\partial_\phi h_{iT\tilde{\theta}\theta}(\hat{\phi}_{iT\tilde{\theta}}(\theta))|}(1 + O_p(T^{-1})),$$

where the $O_p(T^{-1})$ *term holds uniformly in* $\theta \in \Theta$. □

A closed-form expression for $\partial_\phi h_{iT\tilde{\theta}\theta}(\hat{\phi}_{iT\tilde{\theta}}(\theta))$ can be obtained from SST21 (Equation I.4), although it is often easier to obtain $\partial_\phi h_{iT\tilde{\theta}\theta}(\hat{\phi}_{iT\tilde{\theta}}(\theta))$, and $\hat{\phi}_{iT\tilde{\theta}}(\theta)$, directly from the equation defining the inverse ZSE transformation; cf. the panel logit example in Sect. 6.12.3 for details.

6.9 Desirable Properties of the ZSE Transformed IL

The infeasible ZSE transformed IL

$$\bar{L}^0_{iT}(\theta) := \int_{(\mathfrak{a},\mathfrak{b})} \tilde{L}^0_{iT}(\theta,\phi)\pi_i(\phi)\,d\phi = \int_{(\mathfrak{a},\mathfrak{b})} L_{iT}(\theta, h_{iT\theta_0\theta}(\phi))\pi_i(\phi)\,d\phi,$$

and the corresponding log-infeasible-IL $\bar{\ell}^0_{iT}(\theta) := T^{-1}\log\bar{L}^0_{iT}(\theta)$, possess some desirable properties, which is why the MILE is robust to the choice of the weight-functions and behaves very well in finite samples. Specifically,

1. $\bar{L}^0_{iT}(\theta)$ is invariant to interest respecting reparametrizations of $L_{iT}(\theta,\alpha_i)$.
2. The weight-function π_i is irrelevant in the sense that the mean and variance of the IL score are (approximately) independent of π_i.
3. The IL score is first-order unbiased (in some cases, e.g., the static and dynamic Neyman-Scott models in Sects. 6.12.1 and 6.12.5, the score of the IL defined with $\pi_i := 1$ can be exactly unbiased).
4. The information bias is also of order $1/T^2$.

Throughout this section, limits are taken as $T \to \infty$.

6.9.1 *Invariance*

An interest respecting transformation, i.e., a map of the form $(\theta,\alpha_i) \mapsto (\theta, b(\alpha_i))$, where b is a bijection from $(\mathfrak{a},\mathfrak{b}) \to (\mathfrak{a},\mathfrak{b})$, does not change the ZSE transformed likelihood. In other words, $L_{iT}(\theta,\cdot)$ and $L_{iT}(\theta, b^{-1}(\cdot))$ both yield the same $\tilde{L}^0_{iT}(\theta,\cdot)$. Hence, $\bar{L}^0_{iT}(\theta)$ remains invariant to interest respecting transformations. SST21 show that the same argument applies to the feasible integrated likelihood $\bar{L}_{iT}(\theta)$. Consequently, the MILE remains invariant to interesting respecting transformations, a property not shared by the estimators of AB and Lancaster.

Henceforth, the second argument in functionals of $\ell_{iT}(\theta,\alpha_i)$ is omitted when α_i is evaluated at $\alpha^*_{iT}(\theta)$, the population level MLE of α_i. E.g., we write the target log-likelihood as $\ell_{iT}(\theta) := \ell_{iT}(\theta,\alpha^*_{iT}(\theta))$, and $\ell_{iT\alpha}(\theta) := \ell_{iT\alpha}(\theta,\alpha^*_{iT}(\theta))$. Similarly, the second argument in functionals of $\tilde{\ell}^0_{iT}(\theta,\phi)$ is omitted when ϕ is evaluated at α_{i0}, which, by Lemma 6.1, is the population level MLE of ϕ in the ZSE transformed likelihood. E.g., we write $\tilde{\ell}^0_{iT\phi\phi}(\theta) := \tilde{\ell}^0_{iT\phi\phi}(\theta,\phi)|_{\phi=\alpha_{i0}}$.

6.9.2 Irrelevance of the Weight-Function

It can be shown that[6]

$$\bar{\ell}_{iT}^{0}(\theta) = c_T + \ell_{iT}(\theta, \hat{\alpha}_{iT}(\theta)) - \frac{1}{2T}\log(-\mathbb{E}_0 \bar{\ell}_{iT\phi\phi}^{0}(\theta)) - \frac{1}{2T} C_{iT}(\theta)$$
$$+ \frac{1}{T}\log \pi_i(\alpha_{i0}) - \dot{\pi}_i(\alpha_{i0})\frac{1}{T} P_{iT}(\theta) + O_{\mathrm{p}}(\frac{1}{T^2}), \quad \theta \in \Theta, \tag{6.7}$$

where $C_{iT}(\theta)$ defined in SST21 (Equation F.4) satisfies $\mathbb{E}_0 C_{iT}(\theta) = 0$ for each θ, $P_{iT}(\theta)$ defined in SST21 (Equation F.19) satisfies $\mathbb{E}_0 P_{iT}(\theta) = 0$ for each θ (C_{iT} and P_{iT} do not depend on π_i), and $\dot{\pi}_i(\phi) := \partial_\phi \log \pi_i(\phi)$. Let $\bar{\ell}_{iT\theta}^{0}(\theta) := \nabla_\theta \bar{\ell}_{iT}^{0}(\theta)$ denote the score corresponding to $\bar{\ell}_{iT}^{0}(\theta)$. SST21 demonstrate that $\mathbb{E}_0 \bar{\ell}_{iT\theta}^{0}(\theta_0)$ and $\mathrm{var}_0\, \bar{\ell}_{iT\theta}^{0}(\theta_0)$ are approximately independent of π_i in the sense that they do not depend on π_i, up to a term of order $1/T^2$. The fact that the mean and variance of the IL score are independent of the weight-function π_i up to a term of order $1/T^2$, implies that π_i will not affect the limiting distribution of the IL score, hence, that of the MILE, if $n/T^3 \to 0$.

6.9.3 First-Order Score Unbiasedness

SST21 show that the score of the infeasible ZSE transformed IL is first-order unbiased. Namely, for each i, the score of the infeasible ZSE transformed IL satisfies the first Bartlett identity up to a term of order T^{-2}, i.e., $\mathbb{E}_0 \bar{\ell}_{iT\theta}^{0}(\theta_0) = O_{\mathrm{p}}(T^{-2})$.[7]

6.9.4 First-Order Information Unbiasedness

SST21 also show that, for each i, the information equality for the infeasible ZSE transformed IL holds up to an $O_{\mathrm{p}}(T^{-2})$ term, i.e.,[8]

$$T\mathbb{E}_0 \nabla_\theta \bar{\ell}_{iT}^{0}(\theta_0) \partial_\theta \bar{\ell}_{iT}^{0}(\theta_0) + \mathbb{E}_0 \nabla_{\theta\theta}^{2} \bar{\ell}_{iT}^{0}(\theta_0) = O_{\mathrm{p}}(T^{-2}). \tag{6.8}$$

In other words, for each individual, the information of the infeasible ZSE transformed IL is first-order unbiased.

[6] Equation (6.7) uses that the ZSE parameter is strongly unrelated to θ at the population level. Without the strong unrelatedness property, the $O_{\mathrm{p}}(T^{-2})$ term in (6.7) would only be $O_{\mathrm{p}}(T^{-1})$.

[7] This relies on the fact that the ZSE parameter and θ are strongly unrelated at the population level.

[8] The proof of (6.8) also uses the fact that the ZSE parameter and θ are strongly unrelated at the population level.

Information unbiasedness, i.e., the 2nd Bartlett identity, is useful for at least two important reasons. First, information unbiasedness guarantees the asymptotic efficiency of the MLE (which is an M-estimator) by ensuring that the "sandwich form" of its asymptotic variance equals the inverse of the Fisher information. Since the infeasible ZSE transformed IL is always first-order information unbiased (read: close to being information unbiased), whereas the IL of AB or Lancaster is not always first-order information unbiased (read: not close to being information unbiased), this suggests that in finite samples the variance of the MILE may be smaller than the variance of $\hat{\theta}_{\text{AB}}$, at least when T is small. Second, information unbiasedness is useful because it is fundamental for testing model specification. In panel data models with fixed-effects, it is not possible to directly test the specification of $f_{iT| iT,\alpha_{i0};\theta_0}$ because α_{i0} is unobserved. Instead, once the individual-specific nuisance parameters have been eliminated, an information matrix (IM) test can be applied to the ZSE transformed IL $n^{-1}\sum_{i=1}^{n}\bar{\ell}_{iT}(\theta)$. Equation 6.8 suggests that an IM test based on the ZSE transformed IL should have better size properties than an IM test based on the IL of AB or Lancaster, because the information equality is easier to reject using the IL of AB or Lancaster due to the fact that they have higher information bias, namely, of order $O_{\text{p}}(T^{-1})$.

The fact that the score and information of the ZSE transformed IL are both first-order unbiased suggests (as mentioned earlier in Sect. 6.4) that our IL is, at least in an intuitive sense, closer to the target likelihood than the IL of AB or Lancaster.[9]

6.10 Asymptotic Properties of the MILE

The MILE is consistent and asymptotically normal with the limit distribution correctly centered as $n, T_n \to \infty$.

Theorem 6.1 *(Consistency of MILE; SST21, Theorem 8.1)* $\hat{\theta} \xrightarrow{\text{p}} \theta_0$. □

The consistency result does not impose a rate on T_n, i.e., T_n can grow faster or slower than n. SST21 also show that, for each i, information equality holds for the target likelihood, and that $F_{iT} := T\mathbb{E}_0[\nabla_\theta \ell_{iT}(\theta_0)\partial_\theta \ell_{iT}(\theta_0)]$ is the partial information for θ in the presence of α_i, i.e.,

$$T\mathbb{E}_0[\nabla_\theta \ell_{iT}(\theta_0)\partial_\theta \ell_{iT}(\theta_0)] + \mathbb{E}_0\nabla^2_{\theta\theta}\ell_{iT}(\theta_0) = 0$$

$$F_{iT} = F_{iT\theta\theta} - \frac{F'_{iT\alpha\theta}F_{iT\alpha\theta}}{F_{iT\alpha\alpha}},$$

[9] The profile likelihood is, in general, also not first-order information unbiased (DiCiccio et al. 1996, p. 190). For instance, it can be shown that in logit and probit models the information bias of the profile-likelihood is of the order T^{-1}, whereas in the dynamic Neyman-Scott model (Sect. 6.12.5) it is of order T^{-2}. This suggests that our IL is closer to the target likelihood than the profile-likelihood as well.

where $F_{iT\theta\theta} := -\mathbb{E}_0\ell_{iT\theta\theta}$, $F_{iT\alpha\theta} := -\mathbb{E}_0\ell_{iT\alpha\theta}$, and $F_{iT\alpha\alpha} := -\mathbb{E}_0\ell_{iT\alpha\alpha}$. This leads to the following result.

Theorem 6.2 *(Asymptotic normality of MILE; SST21, Theorem 8.2) The MILE is asymptotically normal, i.e.,* $\sqrt{nT_n}(\hat{\theta} - \theta_0) \xrightarrow{\mathrm{d}} \mathrm{N}(0, F^{-1})$ *as* $n/T_n^3 \to 0$, *where* $F := \lim_{n\to\infty} n^{-1} \sum_{i=1}^{n} \mathbb{E}F_{iT_n} = \lim_{T\to\infty} \mathbb{E}F_{1T}$ *is the limit of the average expected partial information for* θ. □

The standard error of $\hat{\theta}$ is easily obtained from the Hessian of the integrated loglikelihood. Letting $\bar{\ell}_{\cdot T}(\theta) := n^{-1} \sum_{i=1}^{n} \bar{\ell}_{iT}(\theta)$ denote the average integrated log-likelihood for the entire sample, it can be shown that $-n^{-1} \sum_{i=1}^{n} \nabla^2_{\theta\theta}\bar{\ell}_{iT_n}(\hat{\theta}) \xrightarrow{\mathrm{p}} F$ as $n, T_n \to \infty$. Hence, the variance of $\hat{\theta}$ is consistently estimated by $\hat{F}^{-1}$ with $\hat{F} := -n^{-1} \sum_{i=1}^{n} \nabla^2_{\theta\theta}\bar{\ell}_{iT_n}(\hat{\theta})$.

6.11 Inference

Since the ZSE transformed IL behaves like the target likelihood, inference for θ_0 can be based on the likelihood ratio (LR) statistic $\mathrm{LR}_{nT}(\theta) := 2nT[\bar{\ell}_{\cdot T}(\hat{\theta}) - \bar{\ell}_{\cdot T}(\theta)]$, constructed using the MILE. Under the conditions maintained in Theorem 6.2,

$$\mathrm{LR}_{nT_n}(\theta_0) \xrightarrow{\mathrm{d}} \chi^2_{\dim(\theta_0)} \qquad \text{as } n, T_n \to \infty. \tag{6.9}$$

This result can be used to test hypotheses and construct confidence regions for θ_0. E.g., (6.9) can be used to show that the lower-level random set $\{\theta \in \Theta : \mathrm{LR}_{nT_n}(\theta) \le k_\tau\}$, where $\tau \in (0, 1)$ and k_τ denotes the $1 - \tau$ quantile of a $\chi^2_{\dim(\theta_0)}$ random variable, is a confidence region for θ_0 whose coverage probability approaches $1 - \tau$ as $n, T_n \to \infty$.

6.12 Examples: The ZSE Approach in Some Familiar Settings

Henceforth, $\Lambda(u) := e^u/(1 + e^u)$, $u \in \mathbb{R}$, is the logistic cdf, $\lambda(u) := d\Lambda(u)/du$ the logistic density, $\mathfrak{N}$ denotes the standard normal cdf, and $\mathfrak{n}(u) := d\mathfrak{N}(u)/du$ the standard normal density. Throughout this section, the ZSE transformed IL is constructed using $\pi_i := 1$ and $(\mathfrak{a}, \mathfrak{b}) = \mathbb{R}$.

6.12.1 Static Neyman-Scott Model

In the Neyman-Scott model (Sect. 6.4.1), the parameter of interest is $\theta := \sigma^2$ and the likelihood for the ith individual is $L_{iT}(\theta, \alpha_i) = (2\pi)^{-T/2}\theta^{-T/2}e^{-\sum_{t=1}^{T}(Y_{it}-\alpha_i)^2/2\theta}$. Hence, $\ell_{iT\alpha}(\theta, \alpha_i) = \sum_{t=1}^{T}(Y_{it} - \alpha_i)/T\theta$ and

$$\mathbb{E}[\ell_{iT\alpha}(\theta, \alpha_i); \check{\theta}, \check{\alpha}] = \frac{1}{T\theta}\sum_{t=1}^{T}(\mathbb{E}[Y_{it}; \check{\theta}, \check{\alpha}] - \alpha_i) = \frac{\check{\alpha} - \alpha_i}{\theta},$$

implying that the ZSE transformation and its inverse are the identity mapping on $\mathbb{R}$, i.e., $g_{iT\theta_0\theta}(\alpha_i) = \alpha_i$ and $h_{iT\theta_0\theta}(\phi) = \phi$. Thus, $\tilde{L}_{iT}(\theta, \phi) = L_{iT}(\theta, \phi)$ because $h_{iT\theta_0\theta}$ does not depend upon θ_0. Hence, letting $\bar{Y}_{i\cdot} := \sum_{t=1}^{T} Y_{it}/T$, the ZSE transformed IL for the ith individual is given by (modulo a factor that does not depend on θ)

$$\begin{aligned}\bar{L}_{iT}(\theta) = \int_{\mathbb{R}} \tilde{L}_{iT}(\theta, \phi)\, d\phi &= \int_{\mathbb{R}} L_{iT}(\theta, \phi)\, d\phi \\ &= (2\pi)^{-(T-1)/2}T^{-1/2}\theta^{-(T-1)/2}e^{-\sum_{t=1}^{T}(Y_{it}-\bar{Y}_{i\cdot})^2/2\theta} \\ &\propto \frac{\text{pdf}_{\ \ iT|\alpha_i;\theta}(\ \ iT)}{\text{pdf}_{\sum_{t=1}^{T} Y_{it}|\alpha_i;\theta}(\sum_{t=1}^{T} Y_{it})}.\end{aligned}$$

Thus, $\bar{L}_{iT}(\theta)$ is equal to the conditional likelihood, which does not depend upon α_i because $\sum_{t=1}^{T} Y_{it}$ is sufficient for α_i. Therefore, the MILE coincides with the conditional maximum likelihood estimator (CMLE), i.e.,

$$\hat{\sigma}^2 = \frac{1}{n(T-1)}\sum_{i=1}^{n}\sum_{t=1}^{T}(Y_{it} - \bar{Y}_{i\cdot})^2.$$

SST21 show that, as $n \to \infty$ but T is held fixed,

$$\begin{aligned}\hat{\sigma}^2 - \sigma_0^2 &= O_{\text{p}}(\sqrt{\frac{1}{nT}}) + O_{\text{p}}(\sqrt{\frac{1}{nT^2}}) \\ \sqrt{n(T-1)}(\hat{\sigma}^2 - \sigma_0^2) &\xrightarrow{\text{d}} \text{N}(0, 2\sigma_0^4).\end{aligned} \tag{6.10}$$

Hence, unlike the fixed-effects MLE, the MILE is consistent as $n \to \infty$, T-fixed. The asymptotic normality of the MILE, as $n \to \infty$ and T is held fixed, is also a much more robust result than the one demonstrated for the fixed-effects MLE in (6.2). Letting $\bar{\ell}_{iT\theta}(\theta) := \nabla_\theta \bar{\ell}_{iT}(\theta)$, note that

$$\bar{\ell}_{iT\theta}(\theta) = -\frac{(T-1)}{T}\frac{1}{2\theta^2}[\theta - \frac{1}{T-1}\sum_{t=1}^{T}(Y_{it} - \bar{Y}_{i\cdot})^2] \implies \mathbb{E}_0\bar{\ell}_{iT\theta}(\theta_0) = 0.$$

Therefore, the individual integrated loglikelihood scores are exactly unbiased for zero. Hence, the MILE is just the generalized method of moments (GMM) estimator of σ^2 based on the moment condition $\mathbb{E}_0[\theta_0 - \frac{1}{T-1}\sum_{t=1}^{T}(Y_{it} - \bar{Y}_{i\cdot})^2] = 0$, and the limiting distribution in (6.10) can also be obtained by applying GMM theory to this moment condition.

6.12.2 Panel Poisson

Let $Y_{it}\big|\ \ _{iT}, \alpha_{i0}; \theta_0 \stackrel{\mathrm{d}}{=} \text{Poisson}(e^{X_{it}'\theta_0+\alpha_{i0}})$, so that the likelihood for the ith individual is

$$L_{iT}(\theta, \alpha_i) = (\prod_{t=1}^{T} Y_{it}!)^{-1} e^{-\sum_{t=1}^{T} e^{X_{it}'\theta+\alpha_i}} e^{\sum_{t=1}^{T} Y_{it}(X_{it}'\theta+\alpha_i)}.$$

Hence, $\ell_{iT\alpha}(\theta, \alpha_i) = T^{-1}(-\sum_{t=1}^{T} e^{X_{it}'\theta+\alpha_i} + \sum_{t=1}^{T} Y_{it})$ and

$$\mathbb{E}[\ell_{iT\alpha}(\theta, \alpha_i); \check{\theta}, \check{\alpha}] = T^{-1}(-\sum_{t=1}^{T} e^{X_{it}'\theta+\alpha_i} + \sum_{t=1}^{T} e^{X_{it}'\check{\theta}+\check{\alpha}}).$$

Consequently, the ZSE transformation and its inverse in the panel poisson model are

$$g_{iT\theta_0\theta}(\alpha_i) = \alpha_i + \log(\frac{\sum_{t=1}^{T} e^{X_{it}'\theta}}{\sum_{t=1}^{T} e^{X_{it}'\theta_0}}) \quad \& \quad h_{iT\theta_0\theta}(\phi) = \phi + \log(\frac{\sum_{t=1}^{T} e^{X_{it}'\theta_0}}{\sum_{t=1}^{T} e^{X_{it}'\theta}}).$$

It follows that $\tilde{L}_{iT}^{0}(\theta, \phi) = C_{iT}(\phi, \theta_0)\dfrac{e^{\sum_{t=1}^{T} Y_{it} X_{it}'\theta}}{(\sum_{t=1}^{T} e^{X_{it}'\theta})^{\sum_{t=1}^{T} Y_{it}}}$, where the constant $C_{iT}(\phi, \theta_0)$ does not depend upon θ. Hence, modulo a factor that does not depend on θ, we have that

$$\bar{L}_{iT}(\theta) = \frac{e^{\sum_{t=1}^{T} Y_{it} X_{it}'\theta}}{(\sum_{t=1}^{T} e^{X_{it}'\theta})^{\sum_{t=1}^{T} Y_{it}}} \propto \text{pmf}_{\ \ iT|\ \ iT, \sum_{t=1}^{T} Y_{it}; \theta}(\ \ iT);$$

i.e., the ZSE transformed IL for individual i is just the conditional density of $\ _{iT}$ given $\ _{iT}$ and the statistic $\sum_{t=1}^{T} Y_{it}$, which is sufficient for α_i. The MILE therefore coincides with the CMLE, which is the fixed-effects MLE for the Poisson family (Lancaster 2002, p. 650).

6.12.3 Panel Logit

Let $Y_{it} = \mathbb{1}(X_{it}'\theta_0 + \alpha_{i0} + U_{it} > 0)$, where

$$U_{i1}, \ldots, U_{iT} \big| \ {}_{iT}, \alpha_{i0} \stackrel{\text{d}}{=} \text{LogisticIID}.$$

Since $\Pr(Y_{it} = y| \ {}_{iT}, \alpha_{i0}; \theta_0) = (\Lambda(X_{it}'\theta_0 + \alpha_{i0}))^y (1 - \Lambda(X_{it}'\theta_0 + \alpha_{i0}))^{1-y} \mathbb{1}_{\{0,1\}}(y)$, $y \in \mathbb{R}$, and the observations are independent across t, the likelihood for the ith individual is $L_{iT}(\theta, \alpha_i) = \prod_{t=1}^T (\Lambda(X_{it}'\theta + \alpha_i))^{Y_{it}} (1 - \Lambda(X_{it}'\theta + \alpha_i))^{1-Y_{it}}$. Consequently, $\ell_{iT\alpha}(\theta, \alpha_i) = T^{-1} \sum_{t=1}^T (Y_{it} - \Lambda(X_{it}'\theta + \alpha_i))$ and

$$\mathbb{E}[\ell_{iT\alpha}(\theta, \alpha_i); \check{\theta}, \check{\alpha}] = T^{-1} \sum_{t=1}^T [\Lambda(X_{it}'\check{\theta} + \check{\alpha}) - \Lambda(X_{it}'\theta + \alpha_i)].$$

Using the MLE $\tilde{\theta}$ as the preliminary estimator, the inverse ZSE transformation $h_{iT\tilde{\theta}\theta}$ solves

$$\sum_{t=1}^T [\Lambda(X_{it}'\tilde{\theta} + \phi) - \Lambda(X_{it}'\theta + h_{iT\tilde{\theta}\theta}(\phi))] = 0, \qquad \phi \in \mathbb{R}. \tag{6.11}$$

Unlike the previous examples, there is no closed form solution for $h_{iT\tilde{\theta}\theta}$. Consequently, there is also no closed form expression for the ZSE transformed IL $\bar{L}_{iT}(\theta) = \int_{\mathbb{R}} (\prod_{t=1}^T (\Lambda(X_{it}'\theta + h_{iT\tilde{\theta}\theta}(\phi)))^{Y_{it}} (1 - \Lambda(X_{it}'\theta + h_{iT\tilde{\theta}\theta}(\phi)))^{1-Y_{it}} d\phi$, which has to be obtained numerically as well.

In this example, we can use Corollary 6.1 to get some intuition behind the form of the ZSE transformed IL. Begin by observing that here $\ell_{iT\alpha\alpha}(\theta, \alpha_i) = -T^{-1} \sum_{t=1}^T \lambda(X_{it}'\theta + \alpha_i)$. Moreover, differentiating (6.11) with respect to ϕ, we get that $\partial_\phi h_{iT\tilde{\theta}\theta}(\phi) = \sum_{t=1}^T \lambda(X_{it}'\tilde{\theta} + \phi) / \sum_{t=1}^T \lambda(X_{it}'\theta + h_{iT\tilde{\theta}\theta}(\phi))$. Hence, as SST21 show that $h_{iT\tilde{\theta}\theta}(\hat{\phi}_{iT\tilde{\theta}}(\theta)) = \hat{\alpha}_{iT}(\theta)$, from Corollary 6.1 we have that

$$\begin{aligned} \bar{L}_{iT}(\theta) = \sqrt{\frac{2\pi}{T}} L_{iT}(\theta, \hat{\alpha}_{iT}(\theta)) (\frac{1}{T} \sum_{t=1}^T \lambda(X_{it}'\theta + \hat{\alpha}_{iT}(\theta)))^{1/2} \\ \times \frac{1}{T^{-1} \sum_{t=1}^T \lambda(X_{it}'\tilde{\theta} + \hat{\phi}_{iT\tilde{\theta}}(\theta))} (1 + O_{\text{p}}(\frac{1}{T})). \end{aligned}$$

It is shown in SST21 (Appendix H) that, for the static panel logit model, $\hat{\phi}_{iT\tilde{\theta}}(\theta)$ is constant in θ, in particular,

$$\hat{\phi}_{iT\tilde{\theta}}(\theta) = \hat{\alpha}_{iT}(\tilde{\theta}), \qquad \theta \in \Theta, \tag{6.12}$$

meaning that strong unrelatedness holds in the panel logit model at the sample level as well. Consequently, the expression for $\bar{L}_{iT}(\theta)$ simplifies to

$$\bar{L}_{iT}(\theta) = M_{iT} L_{iT}(\theta, \hat{\alpha}_{iT}(\theta))(T^{-1} \sum_{t=1}^{T} \lambda(X_{it}'\theta + \hat{\alpha}_{iT}(\theta)))^{1/2}(1 + O_{\mathrm{p}}(T^{-1})), \tag{6.13}$$

where $M_{iT} := \sqrt{2\pi/T}/T^{-1}\sum_{t=1}^{T} \lambda(X_{it}'\tilde{\theta} + \hat{\alpha}_{iT}(\tilde{\theta}))$ does not depend on θ. As noted in Arellano (2003a, p. 443), the leading term in (6.13) is the modified profile-likelihood for panel logit. It is also a saddlepoint approximation to the density of $_{iT}$ given $_{iT}$ and the statistic $\sum_{t=1}^{T} Y_{it}$, which is sufficient for α_i (Levin 1990, p. 278). This suggests that $\hat{\theta}$ is a good approximation of the CMLE. The approximation property of the MILE appears to hold even for small T, although it is obtained using the fixed-effects MLE (which is inconsistent as $n \to \infty$, T-fixed) as the preliminary estimator.

We end this example by explicitly demonstrating that the ZSE parameter is strongly unrelated to θ at the population level. Since $h_{iT\theta_0\theta}$ satisfies (6.11) when $\tilde{\theta}$ is replaced by θ_0, it follows that $\phi_{iT}^{*}(\theta)$ satisfies

$$\begin{aligned}
\sum_{t=1}^{T} \Lambda(X_{it}'\theta_0 + \phi_{iT}^{*}(\theta)) &= \sum_{t=1}^{T} \Lambda(X_{it}'\theta + h_{iT\theta_0\theta}(\phi_{iT}^{*}(\theta))) && (\theta \in \Theta) \\
&= \sum_{t=1}^{T} \Lambda(X_{it}'\theta + \alpha_{iT}^{*}(\theta)) && \text{(Equivariance of optimizers)} \\
&= \sum_{t=1}^{T} \Lambda(X_{it}'\theta_0 + \alpha_{i0}). && \text{(FOC of } \alpha_{iT}^{*}(\theta))
\end{aligned}$$

Hence, $\phi_{iT}^{*}(\theta)$ is constant in θ, with $\phi_{iT}^{*}(\theta) = \phi_{iT}^{*}(\theta_0) = \alpha_{i0}$ for each θ. In other words, the ZSE parameter is strongly unrelated to θ at the population level (compare this with (6.12)).

6.12.4 Panel Probit

Let $Y_{it} = \mathbb{1}(X_{it}'\theta_0 + \alpha_{i0} + U_{it} > 0)$, where

$$U_{i1}, \ldots, U_{iT} \big| \; _{iT}, \alpha_{i0} \overset{\mathrm{d}}{=} \mathrm{NIID}(0, 1).$$

Since the observations are independent across t, the likelihood for the ith individual is $L_{iT}(\theta, \alpha_i) = \prod_{t=1}^{T} (\mathfrak{N}(X_{it}'\theta + \alpha_i))^{Y_{it}} (1 - \mathfrak{N}(X_{it}'\theta + \alpha_i))^{1-Y_{it}}$. Hence,

$$\mathbb{E}[\ell_{iT\alpha}(\theta, \alpha_i); \check{\theta}, \check{\alpha}] = T^{-1} \sum_{t=1}^{T} (\mathfrak{N}(X_{it}'\check{\theta} + \check{\alpha}) - \mathfrak{N}(X_{it}'\theta + \alpha_i))\mathfrak{G}(X_{it}'\theta + \alpha_i),$$

where $\mathfrak{G} := \mathfrak{n}/\mathfrak{N}(1-\mathfrak{N}) > 0$ is the probit weight-function. Therefore, using the fixed-effects MLE as the preliminary estimator, the inverse ZSE transformation $h_{iT\tilde{\theta}\theta}$ solves

$$\sum_{t=1}^{T}(\mathfrak{N}(X_{it}'\tilde{\theta}+\phi)-\mathfrak{N}(X_{it}'\theta+h_{iT\tilde{\theta}\theta}(\phi)))\mathfrak{G}(X_{it}'\theta+h_{iT\tilde{\theta}\theta}(\phi))=0, \qquad \phi\in\mathbb{R}.$$

There is no closed form solution for $h_{iT\tilde{\theta}\theta}$. As in panel logit, $h_{iT\tilde{\theta}\theta}$ and the ZSE transformed IL $\bar{L}_{iT}(\theta)=\int_{\mathbb{R}}(\prod_{t=1}^{T}(\mathfrak{N}(X_{it}'\theta+h_{iT\tilde{\theta}\theta}(\phi)))^{Y_{it}}(1-\mathfrak{N}(X_{it}'\theta+h_{iT\tilde{\theta}\theta}(\phi)))^{1-Y_{it}}\,d\phi$ must be obtained numerically. A simple approximation of $\bar{L}_{iT}(\theta)$ as in (6.13) is not available in this example because $\theta\mapsto\hat{\phi}_{iT\tilde{\theta}}(\theta)$ is not constant for panel probit.

6.12.5 *Dynamic Neyman-Scott Model*

Let $Y_{it}=\mu_0 Y_{i,t-1}+\alpha_{i0}+U_{it}$, $t=1,\ldots,T$, where $\mu_0\in(-1,1)$, Y_{i0} is assumed to be observed (AB, p. 514), and

$$(U_{i1},\ldots,U_{iT})\big|Y_{i0},\alpha_{i0}\overset{\mathrm{d}}{=}\mathrm{NIID}(0,\sigma_0^2).$$

Here, $_{iT}=Y_{i0}$. Hence, the likelihood for the ith individual, conditional on Y_{i0}, is

$$L_{iT}(\mu,\sigma^2,\alpha_i)=(2\pi\sigma^2)^{-T/2}\exp(-\frac{1}{2\sigma^2}\sum_{t=1}^{T}(Y_{it}-\mu Y_{i,t-1}-\alpha_i)^2),$$

which implies that $\ell_{iT\alpha}(\mu,\sigma^2,\alpha_i)=\frac{1}{T}[-\frac{T\alpha_i}{\sigma^2}-\frac{\mu}{\sigma^2}Y_{i0}+\frac{1}{\sigma^2}Y_{iT}+\frac{1-\mu}{\sigma^2}\sum_{t=1}^{T-1}Y_{it}]$.

Hence,

$$\mathbb{E}[\ell_{iT\alpha}(\mu,\sigma^2,\alpha_i);\breve{\mu},\breve{\sigma}^2,\breve{\alpha}]=\frac{1}{T}[-\frac{T\alpha_i}{\sigma^2}-\frac{\mu}{\sigma^2}Y_{i0}+\frac{1}{\sigma^2}\beta_T+\frac{1-\mu}{\sigma^2}\sum_{s=1}^{T-1}\beta_s], \tag{6.14}$$

where $\beta_s:=\int_{\mathbb{R}^T}y_{is}\,\mathrm{pdf}_{\mathrm{N}(\breve{\mu}Y_{i0}+\breve{\alpha},\breve{\sigma}^2)}(y_{i1})\prod_{t=2}^{T}\mathrm{pdf}_{\mathrm{N}(\breve{\mu}y_{i,t-1}+\breve{\alpha},\breve{\sigma}^2)}(y_{it})\,dy_{iT}\ldots dy_{i1}$. SST21 show that

$$\beta_s=\breve{\alpha}\sum_{l=0}^{s-1}\breve{\mu}^l+\breve{\mu}^s Y_{i0}, \qquad s=1,\ldots,T. \tag{6.15}$$

Plugging (6.15) in (6.14), we get

$$\mathbb{E}[\ell_{iT\alpha}(\mu, \sigma^2, \alpha_i); \check{\mu}, \check{\sigma}^2, \check{\alpha}] = \frac{1}{T\sigma^2}[-T\alpha_i - a(\mu, \check{\mu})Y_{i0} + \check{\alpha}c(\mu, \check{\mu})],$$

parameter, follows from where $a(\mu, \check{\mu}) := \mu - \check{\mu}^T - (1-\mu)\frac{\check{\mu}-\check{\mu}^T}{1-\check{\mu}}$, $c(\mu, \check{\mu}) := \frac{1-\check{\mu}^T}{1-\check{\mu}} + (1-\mu)\kappa(\check{\mu})$, and $\kappa(\check{\mu}) := \frac{1}{1-\check{\mu}}(T - \frac{1-\check{\mu}^T}{1-\check{\mu}})$. Thus, letting $\theta := (\mu, \sigma^2)$, the ZSE transformation and its inverse are

$$g_{iT\theta_0\theta}(\alpha_i) = (T\alpha_i + a(\mu, \mu_0)Y_{i0})/c(\mu, \mu_0)$$
$$\& \quad h_{iT\theta_0\theta}(\phi = (\phi c(\mu, \mu_0) - a(\mu, \mu_0)Y_{i0})/T.$$

Using the expression for $h_{iT\theta_0\theta}$, it can be shown that

$$\bar{L}^0_{iT}(\mu, \sigma^2) = \sqrt{T}(\sigma^2)^{-(T-1)/2}\exp(-\frac{1}{2\sigma^2}\sum_{t=1}^{T}(\ddot{Y}_{it} - \mu\ddot{Y}_{i,t-1})^2)/c(\mu, \mu_0), \quad (6.16)$$

where $\ddot{Y}_{it} := Y_{it} - \bar{Y}_{i\cdot}$, $\ddot{Y}_{i,t-1} := Y_{i,t-1} - \bar{Y}_{i,-1}$, and $\bar{Y}_{i,-1} := \sum_{t=1}^{T} Y_{i,t-1}/T$. Replacing μ_0 in (6.16) by the fixed-effects MLE $\tilde{\mu} := \sum_{i=1}^{n}\sum_{t=1}^{T}\ddot{Y}_{it}\ddot{Y}_{i,t-1}/\sum_{i=1}^{n}\sum_{t=1}^{T}\ddot{Y}^2_{i,t-1}$, it follows that the feasible ZSE transformed IL for the ith individual is given by

$$\begin{aligned}\bar{L}_{iT}(\mu, \sigma^2) &= \sqrt{T}(\sigma^2)^{-(T-1)/2}\exp(-\frac{1}{2\sigma^2}\sum_{t=1}^{T}(\ddot{Y}_{it} - \mu\ddot{Y}_{i,t-1})^2)/c(\mu, \tilde{\mu})\\ &= L^{\mathrm{p}}_{iT}(\mu, \sigma^2)\sqrt{T}\sigma/c(\mu, \tilde{\mu}),\end{aligned}$$

because $\hat{\alpha}_{iT}(\mu, \sigma^2) = \bar{Y}_{i\cdot} - \mu\bar{Y}_{i,-1}$ and the profile-likelihood $L^{\mathrm{p}}_{iT}(\mu, \sigma^2) := L_{iT}(\mu, \sigma^2, \hat{\alpha}_{iT}(\mu, \sigma^2)) = (2\pi\sigma^2)^{-T/2}\exp(-\frac{1}{2\sigma^2}\sum_{t=1}^{T}(\ddot{Y}_{it} - \mu\ddot{Y}_{i,t-1})^2)$. Hence, the ZSE transformed IL can be regarded as a "modified" version of the profile-likelihood with the multiplicative correction factor $\sqrt{T}\sigma/c(\mu, \tilde{\mu})$.

The MILE $(\hat{\mu}, \hat{\sigma}^2)$ solves $\nabla_{(\mu,\sigma^2)}\sum_{i=1}^{n}\log\bar{L}_{iT}(\hat{\mu}, \hat{\sigma}^2) = 0$, i.e., $\hat{\mu}$ and $\hat{\sigma}^2$ jointly solve the equations

$$\frac{1}{\hat{\sigma}^2}\sum_{i=1}^{n}\sum_{t=1}^{T}\ddot{Y}_{i,t-1}(\ddot{Y}_{it} - \hat{\mu}\ddot{Y}_{i,t-1}) + \frac{n\kappa(\tilde{\mu})}{c(\hat{\mu}, \tilde{\mu})} = 0$$

$$\hat{\sigma}^2 - \frac{1}{n(T-1)}\sum_{i=1}^{n}\sum_{t=1}^{T}(\ddot{Y}_{it} - \hat{\mu}\ddot{Y}_{i,t-1})^2 = 0.$$

The asymptotic behavior of $\hat{\mu}$ and $\hat{\sigma}^2$ as $n, T \to \infty$ can be obtained from first principles. In particular, SST21 show that $\hat{\mu}$ is a consistent estimator of μ_0 and that $\sqrt{nT_n}(\hat{\mu} - \mu_0) \xrightarrow{d} \mathrm{N}(0, 1 - \mu_0^2)$ as $n, T_n \to \infty$.

References

Andersen, Erling Bernhard (1970). "Asymptotic properties of conditional maximum-likelihood estimators". In: *Journal of the Royal Statistical Society, Series B* 32, pp. 283–301.

Arellano, Manuel (2003a). "Discrete choices with panel data". In: *Investigaciones Econas* XXVII, pp. 423–458.

Arellano, Manuel (2003b). *Panel data econometrics*. Oxford University Press, NewYork, NY, USA.

Arellano, Manuel and Stéphane Bonhomme (2009). "Robust priors in nonlinear panel data models". In: *Econometrica* 77, pp. 489–536.

Arellano, Manuel and Jinyong Hahn (2007). "Understanding bias in nonlinear panel models: Some recent developments". In: *Advances in Economics and Econometrics: Ninth World Congress*. Ed. by R. Blundell, W. Newey, and T. Persson. Vol. 3. Cambridge University Press, Cambridge, UK, pp. 381–409.

Baltagi, Badi H. (1995). *Econometric analysis of panel data*. Wiley, Great Britain.

Bester, C. Alan and Christian Hansen (2009). "A penalty function approach to bias reduction in nonlinear panel models with fixed effects". In: *Journal of Business and Economic Statistics* 27, pp. 131–148.

Carro, Jesus M. (2007). "Estimating dynamic panel data discrete choice models with fixed effects". In: *Journal of Econometrics* 140, pp. 503–528.

Chamberlain, Gary (1980). "Analysis of covariance with qualitative data". In: *Review of Economic Studies* XLVII, pp. 225–238.

Chamberlain, Gary (1982). "Multivariate regression models for panel data". In: *Journal of Econometrics* 18, pp. 5–46.

Chamberlain, Gary (1984). "Panel data". In: *Handbook of Econometrics, vol. II*. Ed. by Z. Griliches and M. D. Intriligator. Elsevier, The Netherlands, pp. 1247–1318.

Chamberlain, Gary (2010). "Binary response models for panel data: Identification and information". In: *Econometrica* 78, pp. 159–168.

Cox, D. R. and N. Reid (1987). "Parameter orthogonality and approximate conditional inference". In: *Journal of the Royal Statistical Society, Series B* 49, pp. 1–39.

De Bin, Riccardo, Nicola Sartori, and Thomas A. Severini (2015). "Integrated likelihoods in models with stratum nuisance parameters". In: *Electronic Journal of Statistics* 9, pp. 1474–1491.

Dhaene, Geert and Koen Jochmans (2015). "Split-panel jackknife estimation of fixed-effect models". In: *Review of Economic Studies* 82, pp. 991–1030.

Dhaene, Geert and Koen Jochmans (2016). "Likelihood inference in an autoregression with fixed effects". In: *Econometric Theory* 32, pp. 1178–1215.

DiCiccio, Thomas J., Michael A. Martin, Steven E. Stern, and G. Alastair Young (1996). "Information bias and adjusted profile likelihoods". In: *Journal of the Royal Statistical Society, Series B* 58, pp. 189–203.

Fernández-Val, Iván (2009). "Fixed effects estimation of structural parameters and marginal effects in panel probit models". In: *Journal of Econometrics* 150, pp. 71–85.

Fernández-Val, Iván and Martin Weidner (2016). "Individual and time effects in nonlinear panel models with large N, T". In: *Journal of Econometrics* 192, pp. 291–312.

Hahn, Jinyong and Guido Kuersteiner (2002). "Asymptotically unbiased inference for a dynamic panel model with fixed effects when both n and T are large". In: *Econometrica* 70, pp. 1639–1657.

Hahn, Jinyong and Guido Kuersteiner (2011). "Bias reduction for dynamic nonlinear panel models with fixed effects". In: *Econometric Theory* 27, pp. 1152–1191.

Hahn, Jinyong and Whitney K. Newey (2004). “Jackknife and analytical bias reduction for nonlinear panel models”. In: *Econometrica* 72, pp. 1295–1319.
Hsiao, Cheng (1986). *Analysis of panel data*. Cambridge University Press, Cambridge, UK.
Lancaster, Tony (2000). “The incidental parameter problem since 1948”. In: *Journal of Econometrics* 95, pp. 391–413.
Lancaster, Tony (2002). “Orthogonal parameters and panel data”. In: *Review of Economic Studies* 69, pp. 647–666.
Levin, Bruce (1990). “The saddlepoint correction in conditional logistic likelihood analysis”. In: *Biometrika* 77, pp. 275–285.
Li, Haihong, Bruce G. Lindsay, and Richard P. Waterman (2003). “Efficiency of projected score methods in rectangular array asymptotics”. In: *Journal of the Royal Statistical Society, Series B* 65, pp. 191–208.
Magnac, Thierry (2004). “Panel binary variables and sufficiency: Generalizing conditional logit”. In: *Econometrica* 72, pp. 1859–1876.
McCullagh, Peter and Robert Tibshirani (1990). “A simple method for the adjustment of profile likelihoods”. In: *Journal of the Royal Statistical Society, Series B* 52, pp. 325–344.
Neyman, Jerzy and Elizabeth L. Scott (1948). “Consistent estimation from partially consistent observations”. In: *Econometrica* 16, pp. 1–32.
Pace, Luigi and Alessandra Salvan (2006). “Adjustments of the profile likelihood from a new perspective”. In: *Journal of Statistical Planning and Inference* 136, pp. 3554–3564.
Pakel, Cavit (2019). “Bias reduction in nonlinear and dynamic panels in the presence of cross-section dependence”. In: *Journal of Econometrics* 213, pp. 459–492.
Schumann, Martin, Thomas A. Severini, and Gautam Tripathi (2021). “Integrated likelihood based inference for nonlinear panel data models with unobserved effects”. In: *Journal of Econometrics* 223, pp. 73–95.
Severini, Thomas A. (2000). *Likelihood methods in statistics*. Oxford University Press, London, UK.
Severini, Thomas A. (2007). “Integrated likelihood functions for non-Bayesian inference”. In: *Biometrika* 94, pp. 529–542.
Wooldridge, Jeffrey M. (2010). *Econometric analysis of cross section and panel data*. 2nd. MIT Press, Cambridge, MA, USA.
Woutersen, Tiemen (2002). “Robustness against incidental parameters”. Manuscript, University of Western Ontario, Department of Economics Research Reports, 2002-8.

Chapter 7
Reducing Score and Information Bias in Panel Data Likelihoods

Abstract This chapter studies the role of information bias in likelihood-based estimation and inference, with a focus on its implications for small-sample performance. Drawing on Schumann et al. (2023), it shows that improvements in the finite-sample behavior of estimators and confidence regions can be traced to first-order information unbiasedness, and explains why this property is particularly important for inference rather than for point estimation. The chapter provides a unifying perspective on several well-known simulation findings in the panel data literature, including the variance-reducing effects of score bias correction, the strong coverage properties of confidence regions based on conditional likelihoods in short panels, and the superior performance of pseudolikelihood methods that are simultaneously first-order score and information unbiased.

Keywords Conditional likelihood · Fixed-effects · Hessian bias · Information bias · Panel data likelihood · Pseudolikelihood · Score bias · Short panels · Small-sample performance

7.1 Introduction

Although it has long been recognized that reducing information bias can improve the small-sample performance of likelihood-based estimators and confidence regions, the literature has offered little explanation for why this occurs. In this chapter, drawing on Schumann et al. (2023), hereafter SST23, we show that the improved performance can be traced to first-order information unbiasedness, and we discuss why this property tends to matter more for inference than for estimation. These insights help clarify several findings from simulation studies in the panel data literature. For example, they shed light on the well-known result that reducing the score bias alone often decreases the finite-sample variance of estimators and improves the coverage accuracy of confidence regions in small samples. They also help explain why confidence regions based on conditional (i.e., conditional on sufficient statistics) likelihoods can exhibit excellent coverage even in very short panels. Moreover, they

M. Taniguchi et al., *Econometrics, Finance, and Time Series Analysis*,
JSS Research Series in Statistics,
https://doi.org/10.1007/978-981-95-8045-3_7

provide a theoretical rationale for the simulation results in SST21 (Schumann et al. 2021), who show that, in panels of short duration, estimators and confidence regions based on pseudolikelihoods that are simultaneously first-order score and information unbiased perform markedly better than those relying only on first-order score unbiasedness. This chapter is intended to provide an expository overview. The detailed technical results, proofs, and simulation findings are in SST23.

7.2 Pseudolikelihood

Let θ denote a (column) vector of parameters of interest. In order to do likelihood-based inference for θ in the presence of nuisance parameters, the typical first step is to eliminate the latter from the joint likelihood of θ and the nuisance parameters. This leads to a "pseudolikelihood" of θ, a function of the data and the parameter of interest with properties similar to those of a likelihood function[1]; typically, pseudolikelihoods are constructed so that they can be used to estimate θ and do inference based on the likelihood ratio statistic. Since nuisance parameters can be eliminated in at least three different ways, e.g., by profiling them out, by conditioning on sufficient statistics, or by integrating them out, how well a resulting pseudolikelihood of θ performs in practice depends on how good an approximation it is to a benchmark likelihood of θ. The benchmark likelihood, which we call the "target" likelihood, is a genuine oracle, i.e., infeasible, likelihood of θ.

The usual approach to determine whether a pseudolikelihood behaves like the target likelihood is to check if it satisfies the 1st Bartlett identity ("score unbiasedness") and the 2nd Bartlett identity ("information unbiasedness"). Since score unbiasedness determines the consistency of likelihood estimators (McCullagh and Tibshirani 1990, Remark 2, p. 329), the literature has focused considerable attention on constructing pseudolikelihoods possessing score bias of smaller order. These include, e.g., marginal and conditional likelihoods (Kalbfleisch and Sprott 1970, 1973; Chamberlain 1980), the modified profile-likelihood (Barndorff 1983), the Cox-Reid approximate conditional likelihood (Cox and Reid 1987), score bias-corrected profile-likelihoods (McCullagh and Tibshirani 1990; Arellano and Hahn 2016), and integrated likelihoods (ILs) (Kalbfleisch and Sprott 1970; Berger et al. 1999; Lancaster 2002; Arellano and Bonhomme 2009). Cf. Severini (2000) and Arellano and Hahn (2007) for additional references.

In contrast, although it is reasonable to believe that likelihood-based methods should lead to even better inference when the information bias is also of smaller order—because when the score and information bias are both small, the pseudolikelihood should, at least in an intuitive sense, be closer to the target likelihood—the existing literature is not very precise about why reducing information bias can lead

[1] A pseudolikelihood of θ is a non-negative random function of θ that does not necessarily satisfy the Bartlett identities. In contrast, a "genuine" likelihood of θ is a non-negative random function of θ that satisfies all of the Bartlett identities (Severini 2000, Sect. 3.5.2).

to improved inference in small samples. For instance, the earliest reference to the potential advantage of reducing information bias that we are aware of is McCullagh and Tibshirani (1990, Sect. 6). However, as they note explicitly on p. 326 of their paper, they have "...no strong argument to support this claim." In other works, this can be inferred from what is not said. E.g., the well-known DiCiccio et al. (1996) paper on reducing information bias never actually states why it should be reduced. Severini (1998, Sect. 4) considers information bias for pseudolikelihoods, again without explicitly saying why having no information bias is beneficial. Mykland (1999, Sect. 3) shows that there is some advantage to having bounded information bias, but he does not show any benefit to reducing the information bias further.

One reason why the aforementioned papers find it difficult to connect information bias reduction with inference may be that they restrict their attention to settings where the source of the randomness comes from a single index, e.g., pure cross-sectional likelihoods based on a large sample and a fixed number of parameters. The lack of variability along a second dimension means that in single index asymptotic expansions where the terms due to information bias appear, there tend to be a number of other components of the same order and, hence, simply reducing information bias does not reduce the order of the terms. In contrast, in a panel data setting where the number of individuals (n) and the length of the panel (T) are both allowed to grow, the presence of a second dimension can be exploited to characterize the effect of reducing information bias on statistical inference. As in Chap. 6, we focus on short panels relevant for microeconometric applications, where both $n, T \to \infty$ such that n grows as least as fast as T.

The idea that simultaneously reducing the score and information bias of panel data pseudolikelihoods can lead to better estimation and inference in small samples is supported by the simulation evidence in SST21. In their paper, SST23 provide a theoretical argument to show why this improved performance can be attributed to first-order score and information unbiasedness. Earlier works in the panel data literature that report improved inference for modified likelihood ratios do not investigate the source of the improvements. SST23 make explicit the connection between the bias of the likelihood ratio and the score and information bias of the pseudolikelihood by showing that the improvement in the bias of the likelihood ratio resulting from reducing the score bias of the pseudolikelihood alone comes exclusively from changing an $O_p(n/T)$ term in its approximation to an $O_p(n/T^3)$ term, and that the improvement from reducing the information bias of the pseudolikelihood alone comes from reducing an $O_p(1/T)$ term to an $O_p(1/T^2)$ term (Lemma 7.1 and its discussion). This can be used to explain why information bias seems to matter more for inference than estimation, and why reducing the pseudolikelihood score bias alone often reduces the variance of the resulting estimator and improves the coverage of likelihood ratio confidence regions in small samples. By connecting the target and conditional (on sufficient statistics) likelihoods (Sect. 7.6), the results in Lemmas 7.1 and 7.2 also apply to conditional likelihoods.

7.3 Notation and Assumptions

We use the notation and assumptions introduced in Chap. 6 and supplement them with additional ones as needed in this chapter. Henceforth, FOSU stands for "first-order score unbiased" and FOIU abbreviates "first-order information unbiased." Let $P_{iT}(\theta)$ denote a pseudolikelihood of θ for individual i after α_i has been eliminated from the joint likelihood $L_{iT}(\theta, \alpha_i)$ by using one of the methods described earlier. The average pseudologlikelihood for the ith individual is $\mathfrak{p}_{iT}(\theta) := T^{-1} \log P_{iT}(\theta)$. The average pseudologlikelihood for the entire sample of individuals, denoted by $\mathfrak{p}_{\cdot T}(\theta) := n^{-1} \sum_{i=1}^{n} \mathfrak{p}_{iT}(\theta)$, can then be used to estimate θ and do inference. Specifically, the estimator of θ obtained by maximizing $\mathfrak{p}_{\cdot T}$ is given by $\hat{\theta}_{\mathfrak{p}} := \operatorname{argmax}_{\theta \in \Theta} \mathfrak{p}_{\cdot T}(\theta)$ (the dependence of $\hat{\theta}_{\mathfrak{p}}$ on n, T is suppressed to keep the notation simple), and the likelihood ratio of $\mathfrak{p}$, as a function of θ, is defined to be $\mathrm{LR}_{nT}^{\mathfrak{p}}(\theta) := 2nT[\mathfrak{p}_{\cdot T}(\hat{\theta}_{\mathfrak{p}}) - \mathfrak{p}_{\cdot T}(\theta)]$.

As in Chap. 6, the common parameter θ_0 is identified as the well-separated global maximizer of the limiting expected target loglikelihood as $n, T \to \infty$. The following is assumed about $\hat{\theta}_{\mathfrak{p}}$.

Assumption 7.1 As $n, T \to \infty$, $\hat{\theta}_{\mathfrak{p}}$ is consistent for θ_0 and

$$\sqrt{nT}(\hat{\theta}_{\mathfrak{p}} - \theta_0) = \begin{cases} O_{\mathrm{p}}(1) + O_{\mathrm{p}}(\frac{1}{\sqrt{T}}) + O_{\mathrm{p}}(\sqrt{\frac{n}{T^3}}) & \text{if } \mathfrak{p} \text{ is FOSU} \\ O_{\mathrm{p}}(1) + O_{\mathrm{p}}(\sqrt{\frac{n}{T}}) & \text{if } \mathfrak{p} \text{ is not FOSU} \\ O_{\mathrm{p}}(1) & \text{if } \mathfrak{p} \text{ is the target loglikelihood.} \end{cases}$$

Consistency of $\hat{\theta}_{\mathfrak{p}}$ can be established under standard regularity conditions; see, e.g., SST21 (Theorem 8.1) for a proof of consistency of the MILE. The first case of Assumption 7.1 applies to the MILE and follows from SST21 (Eqs. G.7 and G.9), while the second case applies to the fixed-effects MLE by SST21 (Equ. C.7). The $O_{\mathrm{p}}(\sqrt{n/T^3})$ term in the first case arises from the fact that $\mathfrak{p}$ is FOSU. The additional $O_{\mathrm{p}}(T^{-1/2})$ term reflects the approximation of $\mathfrak{p}$ by the target likelihood used to establish the asymptotic linearity of $\sqrt{nT}(\hat{\theta}_{\mathfrak{p}} - \theta_0)$. When $\mathfrak{p}$ is not FOSU, the $O_{\mathrm{p}}(\sqrt{n/T^3})$ term is replaced by the larger $O_{\mathrm{p}}(\sqrt{n/T})$ term appearing in the second case. Because the target likelihood is score unbiased, the third case contains neither $O_{\mathrm{p}}(\sqrt{n/T^3})$ nor $O_{\mathrm{p}}(T^{-1/2})$ terms. Moreover, score unbiasedness of the target likelihood implies that if, for each T, the common parameter θ_0 is the unique well-separated zero of the moment condition $\mathbb{E}\nabla_\theta \ell_{iT}(\theta_0) = 0$, then the oracle estimator of θ_0, defined in Sect. 7.6 as the maximizer of the target likelihood over the full sample of individuals, is consistent for θ_0 as $n \to \infty$ with T held fixed.

The following assumption restricts the rates at which n and T can grow.

Assumption 7.2 (i) $\lim_{n,T\to\infty} n/T \in (0, \infty]$, and (ii) $\lim_{n,T\to\infty} n/T^3 = 0$. □

Part (i) of Assumption 7.2 emphasizes that we are modeling short panels by restricting T to grow no faster than n. This ensures that if $\mathfrak{p}_{iT}$ is not FOSU, then we do not simply assume away the resulting bias in $\mathbb{E}_0 \mathrm{LR}^{\mathfrak{p}}_{nT}(\theta_0)$ and $\sqrt{nT}(\hat{\theta}_{\mathfrak{p}} - \theta_0)$ by making T grow faster than n. Part (ii) ensures that $\mathbb{E}_0 \mathrm{LR}^{\mathfrak{p}}_{nT}(\theta_0)$ in (7.9) and (7.10) converges to $\dim(\theta_0)$ as $n, T \to \infty$.

7.4 Score and Information Bias of a Panel Data Pseudolikelihood

Specific panel data pseudolikelihoods considered in this chapter include the profile-likelihood (Sect. 7.5), the conditional likelihood (Sect. 7.7), and the ILs of Lancaster (2002), Arellano and Bonhomme (2009), and SST21. These pseudolikelihoods have different properties because they differ in their construction. E.g., the ILs of Arellano-Bonhomme and SST21 require a preliminary estimator in their construction; namely, Arellano-Bonhomme estimate the expectations appearing in their weight-functions by using time series sample averages, whereas SST21 use the fixed-effects MLE defined in Sect. 7.5 to construct their data-based transformation of the fixed-effects. In contrast, the profile-likelihood, the conditional likelihood, and the IL of Lancaster do not require preliminary estimators.

Recall that $\hat{\theta}_{\mathfrak{p}}$ is assumed to be consistent for θ_0 as $n, T \to \infty$, i.e., $\hat{\theta}_{\mathfrak{p}}$ converges in probability to θ_0 as $n, T \to \infty$. However, depending on the pseudolikelihood, e.g., the profile-likelihood, the distribution of $\hat{\theta}_{\mathfrak{p}}$ may be asymptotically biased in the sense that the limiting distribution of $\sqrt{nT}(\hat{\theta}_{\mathfrak{p}} - \theta_0)$ may not be centered at zero as $n, T \to \infty$. If $\mathfrak{p}$ is such that the distribution of $\hat{\theta}_{\mathfrak{p}}$ is asymptotically unbiased, then the likelihood ratio of $\mathfrak{p}$ evaluated at θ_0, i.e., $\mathrm{LR}^{\mathfrak{p}}_{nT}(\theta_0)$, is distributed as a $\chi^2_{\dim(\theta_0)}$ random variable as $n, T \to \infty$. This result can be used to test hypotheses and construct confidence regions for θ_0. E.g., it can be used to show that the lower-level random set $\{\theta \in \Theta : \mathrm{LR}^{\mathfrak{p}}_{nT}(\theta) \leq k_\tau\}$, where $\tau \in (0, 1)$ and k_τ denotes the $1 - \tau$ quantile of a $\chi^2_{\dim(\theta_0)}$ random variable, is a confidence region for θ_0 whose coverage probability approaches $1 - \tau$ as $n, T \to \infty$. However, if the distribution of $\hat{\theta}_{\mathfrak{p}}$ is asymptotically biased, then the limiting distribution of $\mathrm{LR}^{\mathfrak{p}}_{nT}(\theta_0)$ is no longer $\chi^2_{\dim(\theta_0)}$ and, consequently, the likelihood ratio confidence region has poor empirical coverage even in large samples.

7.4.1 *Score Bias*

A pseudolikelihood for individual i is said to be (exactly) score unbiased if it satisfies the first Bartlett identity: $\mathbb{E}_0 \nabla_\theta \mathfrak{p}_{iT}(\theta_0) = 0$. Here $\mathbb{E}_0$ denotes expectation with respect to $\mathrm{pdf}_{\mathcal{Y}_{1T},\ldots,\mathcal{Y}_{nT}|\mathcal{X}_{1T},\alpha_{10},\ldots,\mathcal{X}_{nT},\alpha_{n0}} = \prod_{i=1}^n f_{\mathcal{Y}_{iT}|\mathcal{X}_{iT},\alpha_{i0};\,\theta_0}$. Score unbiasedness generally does not hold for panel data pseudolikelihoods. Indeed, it is well known that eliminating the individual-specific nuisance parameter from a panel data likelihood creates a bias, due to which it is typically the case that $\mathbb{E}_0 \nabla_\theta \mathfrak{p}_{iT}(\theta_0) = O_{\mathrm{p}}(T^{-1})$ as $T \to \infty$.

Since it is this bias that causes the limiting (as $n, T \to \infty$ at the same rate) distribution of $\sqrt{nT}(\hat{\theta}_\mathrm{p} - \theta_0)$ to be incorrectly centered, a major focus of the literature has been to construct pseudolikelihoods for which the score bias is smaller by an order of magnitude, which ensures that the limiting distribution of $\sqrt{nT}(\hat{\theta}_\mathrm{p} - \theta_0)$ is correctly centered when n grows at least as fast as T, but not too fast, e.g., $n/T^3 \to 0$. Consequently, if a pseudologlikelihood $\mathfrak{p}_{iT}$ satisfies the condition

$$\mathbb{E}_0 \nabla_\theta \mathfrak{p}_{iT}(\theta_0) = O_\mathrm{p}(T^{-2}) \quad \text{as } T \to \infty, \tag{7.1}$$

then $\mathfrak{p}_{iT}$ is said to be FOSU.[2] The left-hand side of (7.1), i.e., $\mathbb{E}_0 \nabla_\theta \mathfrak{p}_{iT}(\theta_0)$, is called the "score bias" of the pseudologlikelihood $\mathfrak{p}_{iT}$.

7.4.2 *Information Bias*

A pseudolikelihood for individual i is said to be (exactly) information unbiased if it satisfies the second Bartlett identity: $T\mathbb{E}_0 \nabla_\theta \mathfrak{p}_{iT}(\theta_0) \partial_\theta \mathfrak{p}_{iT}(\theta_0) + \mathbb{E}_0 \nabla^2_{\theta\theta} \mathfrak{p}_{iT}(\theta_0) = 0$,[3] where $\nabla^2_{ab} := \partial_b \circ \nabla_a$. Eliminating the individual-specific nuisance parameter creates a bias due to which information unbiasedness also does not hold for panel data pseudolikelihoods, and it is typically the case that $T\mathbb{E}_0 \nabla_\theta \mathfrak{p}_{iT}(\theta_0) \partial_\theta \mathfrak{p}_{iT}(\theta_0) + \mathbb{E}_0 \nabla^2_{\theta\theta} \mathfrak{p}_{iT}(\theta_0) = O_\mathrm{p}(T^{-1})$ as $T \to \infty$. Therefore, if a pseudologlikelihood $\mathfrak{p}_{iT}$ can be constructed such that its information bias is smaller by an order of magnitude, i.e., if it satisfies the condition

$$T\mathbb{E}_0 \nabla_\theta \mathfrak{p}_{iT}(\theta_0) \partial_\theta \mathfrak{p}_{iT}(\theta_0) + \mathbb{E}_0 \nabla^2_{\theta\theta} \mathfrak{p}_{iT}(\theta_0) = O_\mathrm{p}(T^{-2}) \quad \text{as } T \to \infty, \tag{7.2}$$

then the pseudologlikelihood $\mathfrak{p}_{iT}$ is said to be FOIU. The left-hand side of (7.2), i.e., $T\mathbb{E}_0 \nabla_\theta \mathfrak{p}_{iT}(\theta_0) \partial_\theta \mathfrak{p}_{iT}(\theta_0) + \mathbb{E}_0 \nabla^2_{\theta\theta} \mathfrak{p}_{iT}(\theta_0)$, is called the "information bias" of the pseudologlikelihood $\mathfrak{p}_{iT}$.

SST23 investigate three mutually exclusive types of pseudolikelihoods: Those that are neither FOSU nor FOIU; those that are FOSU but not FOIU; those that are simultaneously FOSU and FOIU.[4] The most prominent example of a pseudolikelihood that is neither FOSU nor FOIU is the profile-likelihood. Examples of pseudolikelihoods that are FOSU, but not FOIU, are the "information orthogonalizing transformation (IOT)" based IL of Lancaster (2002), and the weighted IL of Arellano and Bonhomme (2009). An example of a pseudolikelihood that is both

[2] If A_{iT} is an array, then the statement $A_{iT} = O_\mathrm{p}(1)$ is understood to hold elementwise.

[3] The factor T before the $\mathbb{E}_0 \nabla_\theta \mathfrak{p}_{iT}(\theta_0) \partial_\theta \mathfrak{p}_{iT}(\theta_0)$ term is not a typo. It is due to the fact that $\mathfrak{p}_{iT}$ is defined to be the average (over T) pseudologlikelihood for individual i.

[4] Pseudolikelihoods that are FOIU but not FOSU can also exist. E.g., SST21 (Supplement, p. 123) have shown that the profile-likelihood in the dynamic Neyman-Scott model (a linear AR(1) panel data model with Gaussian innovations) is FOIU but not FOSU. However, information bias reduction is pointless without score bias reduction. Therefore, we do not consider pseudolikelihoods that are FOIU but not FOSU.

FOSU and FOIU is the IL of SST21, which is based on the "zero-score-expectation (ZSE)" transformation of Severini (2007). Target and conditional likelihoods, being genuine likelihoods, are also FOSU and FOIU.

7.5 A Generic Panel Data Pseudolikelihood

The canonical example of a pseudolikelihood of θ is the profile loglikelihood $\ell^{\mathrm{p}}_{\cdot T}(\theta) := n^{-1}\sum_{i=1}^{n}\ell^{\mathrm{p}}_{iT}(\theta)$, where $\ell^{\mathrm{p}}_{iT}(\theta) := \ell_{iT}(\theta, \hat{\alpha}_{iT}(\theta))$ is the profile loglikelihood for individual i and $\hat{\alpha}_{iT}(\theta) := \operatorname{argmax}_{u\in\operatorname{supp}(\alpha_{i0})}\ell_{iT}(\theta, u)$ denotes the MLE of α_i for a given $\theta \in \Theta$. The fixed-effects MLE of θ, which maximizes the profile-likelihood for the entire sample of individuals, is denoted by $\tilde{\theta} := \operatorname{argmax}_{\theta\in\Theta}\ell^{\mathrm{p}}_{\cdot T}(\theta)$. The profile-likelihood is, in general, neither FOSU nor FOIU (DiCiccio et al., 1996, Sect. 3.1, p. 190).

The fact that the profile-likelihood is not FOSU implies that although the fixed-effects MLE is consistent as $n, T \to \infty$, its distribution is asymptotically biased in the sense that if n and T grow at the same rate, then $\sqrt{nT}(\tilde{\theta} - \theta_0)$ converges in distribution to a Gaussian random vector whose mean is not zero, cf., e.g., Li et al. (2003, Sect. 2) and Hahn and Newey (2004, Theorem 1). Consequently, the profile-likelihood ratio $\mathrm{LR}^{\mathrm{p}}_{nT}(\theta_0) := 2nT[\ell^{\mathrm{p}}_{\cdot T}(\tilde{\theta}) - \ell^{\mathrm{p}}_{\cdot T}(\theta_0)]$ is no longer asymptotically distributed as a $\chi^2_{\dim(\theta_0)}$ random variable. It is, therefore, not surprising that in simulation studies, e.g., SST21 (Sect. 7.10), the profile-likelihood ratio confidence regions are found to have poor coverage in small samples.

Research efforts to solve this problem, namely, to construct a pseudologlikelihood whose maximizer has a correctly centered limiting distribution with variance equal to that of the fixed-effects MLE, have led to several competing candidates. The distinguishing feature of these competitors is that they can all be expressed as an additive correction, explicit or otherwise, to the profile loglikelihood. That is, each pseudologlikelihood $\mathfrak{p}_{iT}$ in the literature that claims to solve the problems besetting the profile loglikelihood can be written as

$$\begin{cases} \ell^{\mathrm{p}}_{iT}(\theta) + C_{iT}(\theta) & \text{if the correction is explicit} \qquad (7.3)\\ \ell^{\mathrm{p}}_{iT}(\theta) + C_{iT}(\theta) + O_{\mathrm{p}}(T^{-2}) & \text{if the correction is implicit,} \qquad (7.4)\end{cases}$$

where $C_{iT}(\theta)$ is the additive correction to $\ell^{\mathrm{p}}_{iT}(\theta)$. Though denoted by the same symbol, the additive corrections in (7.3) and (7.4) can be very different. An explicit additive correction means that one is imposed definitionally, i.e., $\mathfrak{p}_{iT}(\theta) := \ell^{\mathrm{p}}_{iT}(\theta) + C_{iT}(\theta)$; cf., e.g., McCullagh and Tibshirani (1990), Arellano (2003), Sartori (2003), Pace and Salvan (2006), Arellano and Hahn (2007, 2016), Schumann (2023), and the references therein. In contrast, an implicit (or internal) additive correction means that the definition of $\mathfrak{p}_{iT}$ can be used to deduce the existence of a function C_{iT} such that $\mathfrak{p}_{iT}(\theta) - \ell^{\mathrm{p}}_{iT}(\theta) = C_{iT}(\theta) + O_{\mathrm{p}}(T^{-2})$. Examples of pseudologlikelihoods where an additive correction to the profile loglikelihood is implicit include the ILs of Lancaster, Arellano-Bonhomme, and SST21.

Whether corrected explicitly or implicitly, many of the pseudolikelihoods proposed in the literature are only FOSU. E.g., SST21 (Sect. 7.5) have shown that the ILs of Lancaster and Arellano-Bonhomme are not FOIU. Although first-order score unbiasedness is enough to guarantee that the estimators proposed by Lancaster and Arellano-Bonhomme are well behaved as $n, T \to \infty$, simulation results in SST21 (Sect. 7.10) reveal that, in short panels: (i) The empirical coverage of the likelihood ratio confidence regions based on pseudolikelihoods that are simultaneously FOSU and FOIU is much higher than the empirical coverage of the likelihood ratio confidence regions based on pseudolikelihoods that are only FOSU; (ii) Estimators obtained by maximizing pseudolikelihoods that are simultaneously FOSU and FOIU have smaller finite-sample variance than estimators obtained by maximizing pseudolikelihoods that are only FOSU. Sections 7.8 and 7.9 explain why the source of this discrepancy can be attributed to first-order information unbiasedness.

7.6 Target Likelihood

Following Pace and Salvan (2006, Sect. 3.2), the target loglikelihood of θ for the ith individual is $\ell_{iT}(\theta) := \ell_{iT}(\theta, \alpha_{iT}^*(\theta))$, where $\alpha_{iT}^*(\theta) := \operatorname{argmax}_{u \in \operatorname{supp}(\alpha_{i0})} \mathbb{E}_0 \ell_{iT}(\theta, u)$ is the "population level" MLE of α_i for a given θ. The average target loglikelihood for the entire sample of individuals, denoted by $\ell_{\cdot T}(\theta) := n^{-1} \sum_{i=1}^{n} \ell_{iT}(\theta)$, is an oracle, i.e., infeasible, loglikelihood because α_{iT}^* is unknown for each i, T. The target likelihood is a genuine likelihood because $L(\theta, \alpha_{iT}^*(\theta); \mathcal{Y}_{iT}, \mathcal{X}_{iT}) := f_{\mathcal{Y}_{iT}|\mathcal{X}_{iT}, \alpha_{iT}^*(\theta);\theta}(\mathcal{Y}_{iT})$ integrates to one with respect to $\mathcal{Y}_{iT}$ (for each θ, $\mathcal{X}_{iT}$) since $\alpha_{iT}^*(\theta)$ does not depend on $\mathcal{Y}_{iT}$. Consequently, the target loglikelihood $\ell_{iT}(\theta)$ satisfies all of the Bartlett identities. In particular, it satisfies (7.1) and (7.2) exactly, i.e., with the $O_{\mathrm{p}}(T^{-2})$ terms on their right-hand side replaced by zero. Therefore, the target likelihood serves as a natural benchmark to evaluate the performance of other pseudolikelihoods. The oracle estimator of θ_0, which maximizes the target likelihood for the entire sample of individuals, is denoted by $\hat{\theta}^* := \operatorname{argmax}_{\theta \in \Theta} \ell_{\cdot T}(\theta)$, and the target likelihood ratio is $\mathrm{LR}_{nT}^{\text{target}}(\theta_0) := 2nT[\ell_{\cdot T}(\hat{\theta}^*) - \ell_{\cdot T}(\theta_0)]$.

7.7 Conditional Likelihood

In panel data models where sufficient statistics for the fixed-effects exist—e.g., exponential family models such as the Neyman-Scott model, panel logit, panel Poisson, and the panel negative binomial model—the fixed-effects can be eliminated by conditioning on the sufficient statistics, which leads to a conditional likelihood. Maximizing the conditional likelihood yields the conditional maximum likelihood estimator

(CMLE). Like the oracle estimator $\hat{\theta}^*$, the CMLE is consistent and asymptotically unbiased, i.e., its limiting distribution is correctly centered, as $n \to \infty$ and T is held fixed. Since conditional likelihoods are genuine likelihoods (cf. SST23), results for the target likelihood also hold for conditional likelihoods.

7.8 Effect of First-Order Score and Information Unbiasedness of the Pseudolikelihood on the Bias of the Likelihood Ratio

The conditional bias (given the covariates $\mathcal{X}_{1T}, \alpha_{10}, \ldots, \mathcal{X}_{nT}, \alpha_{n0}$) of the likelihood ratio, denoted by $\mathbb{E}_0 \mathrm{LR}^{\mathrm{p}}_{nT}(\theta_0) - \dim(\theta_0)$, is related to the conditional coverage probability of the likelihood ratio confidence region. Specifically, the coverage probability approaches its nominal value as the expectation of the likelihood ratio gets closer to its nominal value.[5] Therefore, we now demonstrate how the score and information bias of $\mathfrak{p}_{iT}$ asymptotically affect $\mathbb{E}_0 \mathrm{LR}^{\mathrm{p}}_{nT}(\theta_0) - \dim(\theta_0)$. To motivate this, we begin by looking at the Neyman and Scott (1948) example considered in Sect. 6.4.1.

7.8.1 Score and Information Bias in the Neyman-Scott Model

The heterogeneous means model of Neyman and Scott provides a nice illustration of the effects of reducing the score and information bias of a pseudolikelihood on the bias of the likelihood ratio. To demonstrate this, we consider the likelihood ratios of three pseudolikelihoods and, for the sake of comparison, that of the target likelihood. The three pseudologlikelihoods we consider are the profile loglikelihood (ℓ^{p}_{iT}), the integrated loglikelihood of SST21 ($\bar{\ell}^{\mathrm{zse}}_{iT}$), and the explicitly corrected (or adjusted) pseudologlikelihood $\mathfrak{p}^{\mathrm{adj}}_{iT}(\theta) := \ell^{\mathrm{p}}_{iT}(\theta) + C_{iT}(\theta)$ with $C_{iT}(\theta) := -\dfrac{T^{-1}\sum_{t=1}^{T} \ddot{Y}^2_{it}}{2(T-1)\theta}$. SST23 show that in the Neyman-Scott model, $\bar{\ell}^{\mathrm{zse}}_{iT}$ is score as well as information unbiased, $\mathfrak{p}^{\mathrm{adj}}_{iT}$ is score unbiased but not FOIU, and that ℓ^{p}_{iT} is neither FOSU nor FOIU. Motivated by the arguments in Schumann (2023), the correction term in $\mathfrak{p}^{\mathrm{adj}}_{iT}$ is chosen such that the global maximizer of $\mathfrak{p}^{\mathrm{adj}}_{\cdot T}$ coincides with the global maximizer of $\bar{\ell}^{\mathrm{zse}}_{\cdot T}$ (cf. the proof of (7.6)). This ensures that any differences between the

[5] For some intuition, suppose that $n = 1$ and there are no nuisance parameters and covariates. Let $\mathrm{LR}_T := \mathrm{LR}_T(\theta_0)$; its expectation $\mu_T := \mathbb{E}\mathrm{LR}_T$ now only depends on T. It is known from the theory of Bartlett correction of the likelihood ratio (Barndorff-Nielsen and Hall 1988) that $\mu_T = \dim(\theta_0) + O(T^{-1})$ and $\Pr(\dim(\theta_0)\mathrm{LR}_T/\mu_T \le u) = F(u) + O(T^{-2})$, $u \in \mathbb{R}$, as $T \to \infty$, where F is the cumulative distribution function (c.d.f.) of a $\chi^2_{\dim(\theta_0)}$ random variable. Hence, $\Pr(\mathrm{LR}_T \le u) = F(u) + O(\mu_T - \dim(\theta_0)) + O(T^{-2})$. Therefore, $\mu_T - \dim(\theta_0)$, the bias of the likelihood ratio, helps determine the accuracy of the χ^2-approximation to the c.d.f. of the likelihood ratio. This suggests that the same also holds for panel data pseudolikelihoods.

likelihood ratios of $\mathfrak{p}^{\text{adj}}_{\cdot T}$ and $\bar{\ell}^{\text{zse}}_{\cdot T}$ can be attributed solely to the differences between the pseudolikelihoods themselves and not to the estimators used to construct the likelihood ratios. The bias of the likelihood ratio in the Neyman-Scott model can be decomposed as follows (the limits in (7.5)–(7.8) are taken as $n \to \infty$, T-fixed).[6]

$$\mathbb{E}_0 \mathrm{LR}^{\mathfrak{p}}_{nT}(\theta_0) - 1 = \begin{cases} O(\frac{1}{nT}) + \underbrace{0}_{\text{due to score bias}} + \underbrace{0}_{\text{due to info bias}} & \text{if } \mathfrak{p} \text{ is } \bar{\ell}^{\text{zse}}_{\cdot T} \quad (7.5) \\ O(\frac{1}{nT}) + \underbrace{0}_{\text{due to score bias}} + \underbrace{O(\frac{1}{T})}_{\text{due to info bias}} & \text{if } \mathfrak{p} \text{ is } \mathfrak{p}^{\text{adj}}_{\cdot T} \quad (7.6) \\ \underbrace{O(\frac{n}{T})}_{\text{due to score bias}} + \underbrace{O(\frac{1}{T})}_{\text{due to info bias}} & \text{if } \mathfrak{p} \text{ is } \ell^{\mathrm{p}}_{\cdot T} \quad (7.7) \\ O(\frac{1}{nT}) + \underbrace{0}_{\text{due to score bias}} + \underbrace{0}_{\text{due to info bias}} & \text{if } \mathfrak{p} \text{ is target likelihood.} \quad (7.8) \end{cases}$$

Equations (7.5) and (7.8) show that, asymptotically, the likelihood ratio of $\bar{\ell}^{\text{zse}}_{\cdot T}$ has the same bias as the likelihood ratio of the target likelihood, and the bias converges to zero as $n \to \infty$ and T is held fixed. In contrast, (7.6) and (7.7) reveal that the bias of the likelihood ratio of $\mathfrak{p}^{\text{adj}}_{\cdot T}$ converges to zero when T is allowed to grow with n, and the profile-likelihood ratio bias vanishes only if T grows faster than n, which is not the right setting for modeling short panels. For $\mathfrak{p} \in \{\bar{\ell}^{\text{zse}}_{\cdot T}, \mathfrak{p}^{\text{adj}}_{\cdot T}\}$, the likelihood ratio bias is smallest when $\mathfrak{p} = \bar{\ell}^{\text{zse}}_{\cdot T}$. As $\bar{\ell}^{\text{zse}}_{\cdot T}$ and $\mathfrak{p}^{\text{adj}}_{\cdot T}$ are both score unbiased with the same global maximizer, this can be attributed to the fact that $\bar{\ell}^{\text{zse}}_{\cdot T}$ is information unbiased whereas $\mathfrak{p}^{\text{adj}}_{\cdot T}$ is not. Indeed, their proofs reveal that the $O(1/T)$ term in (7.6), which is absent in (7.5), arises because $\mathfrak{p}^{\text{adj}}_{\cdot T}$ is not FOIU. The $O(n/T)$ term in (7.7), which is due to the fact that the profile-likelihood is not FOSU, reveals that the bias of the profile-likelihood ratio can be large when T is small, even if n is large. In (7.8), there are no terms due to score and information bias because the target likelihood is a genuine likelihood and satisfies all of the Bartlett identities.

The results for the Neyman-Scott model can be generalized considerably to include nonlinear panel data models. The following proposition shows that, asymptotically, the bias of the likelihood ratio of a pseudologlikelihood $\mathfrak{p}_{iT}$ can be decomposed as the sum of two terms: One caused by the score bias of $\mathfrak{p}_{iT}$, and the other caused by the information bias of $\mathfrak{p}_{iT}$.

Lemma 7.1 (Bias of the likelihood ratio for nonlinear panel data models; SST23, Lemma 6.1) *As $n, T \to \infty$,*

[6] In the Neyman-Scott model, the MILE of SST21 and the oracle estimator are consistent for θ_0 as $n \to \infty$ and T is held fixed, whereas the fixed-effects MLE, which maximizes the profile-likelihood, is consistent for θ_0 only when both $n, T \to \infty$. Furthermore, as the only explanatory variables in the Neyman-Scott model are the fixed-effects, $\mathbb{E}_0 \mathrm{LR}^{\mathfrak{p}}_{nT}(\theta_0)$ denotes expectation with respect to $\prod_{i=1}^{n} f_{y_{iT}|x_{iT},\alpha_{i0};\theta_0} = \prod_{i=1}^{n} \prod_{t=1}^{T} \mathrm{pdf}_{\mathrm{N}(0,\sigma_0^2)}$.

$$\mathbb{E}_0 \mathrm{LR}^{\mathfrak{p}}_{nT}(\theta_0) - dim(\theta_0) = \begin{cases} \underbrace{O_{\mathrm{P}}(\frac{n}{T^3})}_{\text{due to score bias}} + \underbrace{O_{\mathrm{P}}(\frac{1}{T^2})}_{\text{due to info bias}} & \text{if } \mathfrak{p} \text{ is FOSU and FOIU} \quad (7.9) \\ \underbrace{O_{\mathrm{P}}(\frac{n}{T^3})}_{\text{due to score bias}} + \underbrace{O_{\mathrm{P}}(\frac{1}{T})}_{\text{due to info bias}} & \text{if } \mathfrak{p} \text{ is FOSU but not FOIU} \quad (7.10) \\ \underbrace{O_{\mathrm{P}}(\frac{n}{T})}_{\text{due to score bias}} + \underbrace{O_{\mathrm{P}}(\frac{1}{T})}_{\text{due to info bias}} & \text{if } \mathfrak{p} \text{ is neither FOSU nor FOIU} \quad (7.11) \\ O_{\mathrm{P}}(\frac{1}{nT}) \text{ if } \mathfrak{p} \text{ is the target loglikelihood.} & \quad (7.12) \end{cases}$$

The limits in Lemma 7.1 are taken as $n, T \to \infty$ because estimators in nonlinear panel data models are generally only consistent as $n, T \to \infty$. The results in (7.9)–(7.12) are qualitatively very similar to those in (7.5)–(7.8), even though we are no longer in the Neyman-Scott setting. The terms due to the score bias depend on both n and T, whereas the terms due to the information bias depend only on T. The $O_{\mathrm{P}}(T^{-2})$ term appears on the right-hand side only when the pseudolikelihood is FOIU.[7] If the pseudolikelihood is not FOIU, then the $O_{\mathrm{P}}(T^{-2})$ term becomes an $O_{\mathrm{P}}(T^{-1})$ term. The $O_{\mathrm{P}}(n/T^3)$ term arises if the pseudolikelihood is FOSU.[8] If the pseudolikelihood is not FOSU, then the $O_{\mathrm{P}}(n/T^3)$ term becomes $O_{\mathrm{P}}(n/T)$. The $O_{\mathrm{P}}(n/T)$ term in (7.11) does not vanish because T does not grow faster than n (Assumption 7.2(i)). This explains why the profile-likelihood ratio behaves very poorly in nonlinear panel data models when T is small, even if n happens to be large.

Lemma 7.1 can be used to justify the improvements that researchers often discover in their simulation experiments when they compare the coverage probability of a likelihood ratio confidence region based on a FOSU pseudolikelihood with one based on the profile-likelihood. Indeed, the improvement in going from (7.11) to (7.10) is caused by the $O_{\mathrm{P}}(n/T)$ term in (7.11) becoming the $O_{\mathrm{P}}(n/T^3)$ term in (7.10). Similarly, comparing the $O_{\mathrm{P}}(1/T^2)$ term in (7.9) with the $O_{\mathrm{P}}(1/T)$ term in (7.10) suggests that the coverage probability of a confidence region obtained by inverting the likelihood ratio of a pseudolikelihood that is both FOSU and FOIU is likely to be higher than the coverage probability of a confidence region based on

---

[7] There is no $O_{\mathrm{P}}(T^{-2})$ term in (7.5) because $\bar{\ell}^{\mathrm{zse}}_{\cdot T}$ is information unbiased in the Neyman-Scott example in Sect. 7.8.1.

[8] There is no $O_{\mathrm{P}}(n/T^3)$ term in (7.5) and (7.6) because, in the Neyman-Scott example in Sect. 7.8.1, $\bar{\ell}^{\mathrm{zse}}_{\cdot T}$ and $\mathfrak{p}^{\mathrm{adj}}_{\cdot T}$ are score unbiased. Unlike (7.5) and (7.6), no $O_{\mathrm{P}}(1/nT)$ term appears in (7.9) and (7.10) because it is dominated by the $O_{\mathrm{P}}(n/T^3)$ term.

a pseudolikelihood that is only FOSU. Therefore, it makes sense to base inference on a pseudolikelihood that is both FOSU and FOIU.

Since the terms due to the score bias depend on both n and T, whereas the terms due to the information bias depend only on T, increasing n alone affects the score bias terms but not the information bias terms. If n and T grow at the same rate, an assumption maintained in several papers, then $O_{\mathrm{P}}(n/T^3) = O_{\mathrm{P}}(1/T^2)$ and the improvement to the coverage probability in (7.9) from first-order information unbiasedness should be comparable to the improvement from first-order score unbiasedness. But if n grows too fast, say, $n = O(T^2)$ (so that the $O_{\mathrm{P}}(n/T^3)$ term still vanishes asymptotically), then the $O_{\mathrm{P}}(n/T^3)$ term in (7.9) becomes $O_{\mathrm{P}}(1/T)$, and an improvement to the coverage probability from first-order information unbiasedness is dominated by the improvement from first-order score unbiasedness. However, the simulation evidence in SST21 suggests that improvements to coverage probability in finite samples exist even when n is much larger than T. Therefore, using a pseudolikelihood that is simultaneously FOSU and FOIU is always beneficial.

Remark 7.1 (i) Lemma 7.1 makes it clear that there is no gain from first-order information unbiasedness without first-order score unbiasedness. Indeed, if $\mathfrak{p}$ were FOIU but not FOSU, then the $O_{\mathrm{P}}(1/T)$ on the right-hand side of (7.11) would be replaced by an $O_{\mathrm{P}}(1/T^2)$ term, i.e.,

$$\mathbb{E}_0 \mathrm{LR}^{\mathfrak{p}}_{nT}(\theta_0) - \dim(\theta_0)\big|_{\mathfrak{p}\text{ is FOIU but not FOSU}} = \underbrace{O_{\mathrm{P}}(\frac{n}{T})}_{\text{due to score bias}} + \underbrace{O_{\mathrm{P}}(\frac{1}{T^2})}_{\text{due to info bias}} = O_{\mathrm{P}}(\frac{n}{T}).$$

(ii) Since conditional likelihoods are genuine likelihoods, the proof of (7.12) can be adapted to show that it also holds for conditional likelihoods (cf. (iii) below). This explains why the likelihood ratio confidence regions based on conditional likelihoods can have excellent coverage probabilities even in very short panels.

(iii) For panel Poisson, Lancaster (2002, p. 650) showed that the profile-likelihood for individual i coincides with the conditional density of $\mathcal{Y}_{iT}$ given $\mathcal{X}_{iT}$ and the statistic $\sum_{t=1}^{T} Y_{it}$, which is sufficient for α_i. As shown in Sect. 6.12.2, the same also holds for the ZSE transformed IL of SST21. Therefore, in the panel Poisson model, the profile-likelihood, the ZSE transformed IL, and the conditional likelihood all coincide and, as noted in Remark 7.1(ii), satisfy (7.12). □

7.9 Effect of First-Order Score and Information Unbiasedness of the Pseudolikelihood on the Bias and Variance of the Estimator

The following result shows how $\mathfrak{p}$ being FOSU or FOIU affects the conditional (on the covariates $\mathcal{X}_{1T}, \alpha_{10}, \ldots, \mathcal{X}_{nT}, \alpha_{1n}$) bias and variance of $\hat{\theta}_{\mathfrak{p}}$.

Lemma 7.2 (Bias and variance of the estimator for nonlinear panel data models; SST23, Lemma 6.2) *As* $n, T \to \infty$,

$$\mathbb{E}_0(\hat{\theta}_p) - \theta_0 = \begin{cases} \underbrace{O_{\mathrm{P}}(\frac{1}{T^2})}_{\text{due to score bias}} & \textit{if } \mathfrak{p} \textit{ is FOSU} \quad (7.13) \\ \underbrace{O_{\mathrm{P}}(\frac{1}{T})}_{\text{due to score bias}} & \textit{if } \mathfrak{p} \textit{ is not FOSU} \quad (7.14) \\ O_{\mathrm{P}}(\frac{1}{nT}) & \textit{if } \mathfrak{p} \textit{ is the target loglikelihood.} \quad (7.15) \end{cases}$$

Furthermore, if the expected Hessian $H_{\mathfrak{p}.T} := n^{-1}\sum_{i=1}^{n}\mathbb{E}_0\nabla^2_{\theta\theta}\mathfrak{p}_{iT}(\theta_0) =: n^{-1}\sum_{i=1}^{n}H_{\mathfrak{p}iT}$ *and* var_0 *is the variance using* $\mathbb{E}_0$*, then*

$$\mathrm{var}_0\sqrt{nT}(\hat{\theta}_p - \theta_0) + H_{\mathfrak{p}.T}^{-1} = \begin{cases} \underbrace{O_{\mathrm{P}}(\frac{n}{T^3})}_{\text{due to score bias}} + \underbrace{O_{\mathrm{P}}(\frac{1}{T^2})}_{\text{due to info bias}} & \textit{if } \mathfrak{p} \textit{ is FOSU and FOIU} \quad (7.16) \\ \underbrace{O_{\mathrm{P}}(\frac{n}{T^3})}_{\text{due to score bias}} + \underbrace{O_{\mathrm{P}}(\frac{1}{T})}_{\text{due to info bias}} & \textit{if } \mathfrak{p} \textit{ is FOSU but not FOIU} \quad (7.17) \\ \underbrace{O_{\mathrm{P}}(\frac{n}{T})}_{\text{due to score bias}} + \underbrace{O_{\mathrm{P}}(\frac{1}{T})}_{\text{due to info bias}} & \textit{if } \mathfrak{p} \textit{ is neither FOSU nor FOIU} \quad (7.18) \\ O_{\mathrm{P}}(\frac{1}{nT}) & \textit{if } \mathfrak{p} \textit{ is the target loglikelihood.} \quad (7.19) \end{cases}$$

Lemma 7.2 reveals some interesting findings. For instance, (7.13) and (7.14) show that, asymptotically, the bias of $\hat{\theta}_{\mathfrak{p}}$ depends only on T, and that its magnitude is determined solely by whether $\mathfrak{p}$ is FOSU or not. In particular, $\mathfrak{p}$ being FOIU does not reduce the bias of $\hat{\theta}_{\mathfrak{p}}$. Equation (7.15) shows that the bias of the oracle estimator goes to zero as $n \to \infty$ and T is held fixed, which is not surprising because the oracle estimator can be consistent for θ_0 as $n \to \infty$ and T is held fixed.

To assess the effect of first-order score and information unbiasedness of $\mathfrak{p}$ on the variance of estimators, we rewrite the inverse Hessian matrix $H_{\mathfrak{p}.T}^{-1}$ in (7.16)–(7.19) in order to facilitate their comparison for different pseudologlikelihoods. Indeed, since the asymptotic variance of $\hat{\theta}_{\mathfrak{p}}$ is equal to the asymptotic variance of the oracle estimator, it is reasonable to express $-H_{\mathfrak{p}iT}$ as a deviation from $F_{iT} := -H_{\mathfrak{p}iT}|_{\mathfrak{p}=\text{target}} = -\mathbb{E}_0\nabla^2_{\theta\theta}\ell_{iT}(\theta_0, \alpha^*_{iT}(\theta_0))$, the Fisher information of the target likelihood for individual i.[9] To derive an expression for $H_{\mathfrak{p}_{iT}}$ in terms of its deviation from F_{iT}, recall from Sect. 7.5 that any panel data pseudologlikelihood $\mathfrak{p}_{iT}$ that claims to better the profile loglikelihood can be written as

$$\mathfrak{p}_{iT}(\theta) = \ell^{\mathrm{P}}_{iT}(\theta) + C_{iT}(\theta) + O_{\mathrm{p}}(T^{-2}) \quad \text{as } T \to \infty, \tag{7.20}$$

where C_{iT} is a correction term designed to improve specific properties of the profile loglikelihood (if the correction is explicit, then the $O_{\mathrm{p}}(T^{-2})$ term in (7.20) is identically zero). SST23 show that if $\mathbb{E}_0 \nabla^2_\theta C_{iT}(\theta_0) = O_{\mathrm{p}}(T^{-1})$, then

$$-H_{\mathfrak{p}_{iT}} = F_{iT} + \underbrace{O_{\mathrm{p}}(T^{-1})}_{\text{Hessian bias}} \quad \text{as } T \to \infty, \tag{7.21}$$

where "Hessian bias" of a pseudolikelihood $\mathfrak{p}_{iT}$ is defined to be $-H_{\mathfrak{p}_{iT}} - F_{iT}$. Therefore, analogous to the definitions of first-order score and information unbiasedness in (7.1) and (7.2), we call a pseudolikelihood $\mathfrak{p}_{iT}$ to be "first-order Hessian unbiased" (FOHU) whenever

$$-H_{\mathfrak{p}_{iT}} = F_{iT} + O_{\mathrm{p}}(T^{-2}) \quad \text{as } T \to \infty \text{ is the target loglikelihood.} \tag{7.22}$$

Since the target likelihood satisfies (7.22) exactly, i.e., with the $O_{\mathrm{p}}(T^{-2})$ terms identically zero, the target likelihood is, by definition, Hessian unbiased.

Letting $F_{\cdot T} := n^{-1}\sum_{i=1}^{n} F_{iT}$ denote average Fisher information of the target likelihood for the entire sample, we have $-H^{-1}_{\mathfrak{p}\cdot T} \overset{(7.21)}{=} F^{-1}_{\cdot T} + O_{\mathrm{p}}(T^{-1})$. Hence, (7.16), (7.17), and (7.18) imply that the difference $\mathrm{var}_0 \sqrt{nT}(\hat{\theta}_{\mathfrak{p}} - \theta_0) - F^{-1}_{\cdot T}$ is equal to

$$\begin{cases} \underbrace{O_{\mathrm{p}}(\frac{1}{T})}_{\text{due to Hessian bias}} + \underbrace{O_{\mathrm{p}}(\frac{n}{T^3})}_{\text{duetoscorebias}} + \underbrace{O_{\mathrm{p}}(\frac{1}{T^2})}_{\text{due to info bias}} & \text{if } \mathfrak{p} \text{ is FOSU and FOIU} & (7.23) \\ \underbrace{O_{\mathrm{p}}(\frac{1}{T})}_{\text{due to Hessian bias}} + \underbrace{O_{\mathrm{p}}(\frac{n}{T^3})}_{\text{due to score bias}} + \underbrace{O_{\mathrm{p}}(\frac{1}{T})}_{\text{due to info bias}} & \text{if } \mathfrak{p} \text{ is FOSU but not FOIU} & (7.24) \\ \underbrace{O_{\mathrm{p}}(\frac{1}{T})}_{\text{due to Hessian bias}} + \underbrace{O_{\mathrm{p}}(\frac{n}{T})}_{\text{due to score bias}} + \underbrace{O_{\mathrm{p}}(\frac{1}{T})}_{\text{due to info bias}} & \text{if } \mathfrak{p} \text{ is neither FOSU nor FOIU.} & (7.25) \end{cases}$$

A well-documented phenomenon in the panel data literature—cf., e.g., the simulation results in Hahn and Newey (2004), Fernández-Val (2009), and SST21—is that in simulations of fixed-effects logit and probit models, estimators that correct the bias of the pseudolikelihood score not only perform better in

[9] The Fisher information, defined as the variance of the score function, is equal to the negative of the expected Hessian because the target likelihood satisfies the first two Bartlett identities.

terms of bias, but also have a lower finite-sample variance than the fixed-effects MLE. One explanation for this finding is provided by Hahn and Newey (2004, p. 1307), who state that if the score bias can be attributed to the (normalized) scale parameter, then reducing the bias may lead to a decrease in the variance. An alternative explanation is given by (7.25) and (7.24): Reducing the score bias causes the $O_{\mathrm{p}}(n/T)$ term in (7.25) to be replaced by the $O_{\mathrm{p}}(n/T^3)$ term in (7.24), thereby reducing the variance.

In the decomposition for $\mathrm{var}_0\sqrt{nT}(\hat{\theta}_{\mathfrak{p}}-\theta_0)-F_{\cdot T}^{-1}$, the term due to the Hessian bias of $\mathfrak{p}$ dominates the term due to the information bias of $\mathfrak{p}$ by an order of magnitude when $\mathfrak{p}$ is FOIU. In contrast, the Hessian bias of a pseudologlikelihood does not enter (7.9)–(7.12) at all. This explains why first-order information unbiasedness of a pseudologlikelihood has a greater impact on inference than on estimation. Unlike for the estimation bias, (7.23), (7.24), and (7.25) suggest that $\mathfrak{p}$ being FOIU can reduce the deviation of the variance of $\hat{\theta}_{\mathfrak{p}}$ from the variance of the oracle estimator: The deviation is the largest if $\mathfrak{p}$ is neither FOSU nor FOIU, smaller if $\mathfrak{p}$ is FOSU but not FOIU, and the smallest if $\mathfrak{p}$ is both FOSU and FOIU. Since this ranking is based on comparing the rates inside the O_{p} terms in (7.23), (7.24), and (7.25), it may not hold if the constants in the O_{p} terms are not comparable across the pseudolikelihoods.

The insights from (7.23) and (7.24) also apply to conditional (on sufficient statistics) likelihoods. They reveal that the conditional (on the covariates) mean squared error (mse_0) of the CMLE in small samples may be comparable to—instead of being smaller than—the MSE of $\hat{\theta}_{\mathfrak{p}}$ if $\mathfrak{p}$ is FOSU.[10] This suggests that bias-corrected estimators may be viable alternatives for CMLEs even in models where the latter exist. Indeed, although conditional likelihoods are score and information unbiased because they are genuine likelihoods (Remark 7.7), they are not FOHU in general.[11] Therefore, since there are no O_{p} terms due to the score and information bias, $\mathrm{var}_0\sqrt{nT}(\hat{\theta}_{\mathrm{CMLE}}-\theta_0)\overset{(7.23)}{=}F_{\cdot T}^{-1}+O_{\mathrm{p}}(T^{-1})$, where the $O_{\mathrm{p}}(T^{-1})$ term is due to the Hessian bias alone. By (7.15), the bias of the CMLE is $O_{\mathrm{p}}(1/nT)$. Hence,

$$\mathrm{mse}_0(\hat{\theta}_{\mathrm{CMLE}})=\frac{1}{nT}[F_{\cdot T}^{-1}+O_{\mathrm{p}}(\frac{1}{T})]+O_{\mathrm{p}}(\frac{1}{n^2T^2}).$$

Similarly, by (7.13) and (7.24),

$$\mathrm{mse}_0(\hat{\theta}_{\mathfrak{p}})=\frac{1}{nT}[F_{\cdot T}^{-1}+O_{\mathrm{p}}(\frac{1}{T})+O_{\mathrm{p}}(\frac{n}{T^3})+O_{\mathrm{p}}(\frac{1}{T})]+O_{\mathrm{p}}(\frac{1}{T^4}).$$

Therefore, $\mathrm{mse}_0(\hat{\theta}_{\mathrm{CMLE}})=\mathrm{mse}_0(\hat{\theta}_{\mathfrak{p}})$ if n, T grow at the same rate.

In the decomposition for $\mathrm{var}_0\sqrt{nT}(\hat{\theta}_{\mathfrak{p}}-\theta_0)-F_{\cdot T}^{-1}$ given in (7.23), (7.24), and (7.25), the Hessian bias of $\mathfrak{p}$ is of smaller order than its information bias, even when $\mathfrak{p}$ is FOIU. This naturally suggests looking for an additive adjustment to the profile loglikelihood that yields a pseudologlikelihood which, in addition to being FOSU, is also FOIU and has its Hessian bias reduced by an additional order of magnitude. Equivalently, one may ask: Does there exist a FOSU pseudolikelihood $\mathfrak{p}_{iT}$ that is simultaneously FOIU and FOHU? Apart from the target likelihood—which is score, information, Hessian-unbiased by definition—it is straightforward to show that the answer is no in the Neyman-Scott model (for the pseudolikelihoods in Sect. 7.8.1, $\bar{\ell}_{iT}^{\mathrm{ZSE}}$ is {score, information}-unbiased but not FOHU, whereas $\mathfrak{p}_{iT}^{\mathrm{ADJ}}$ is {score, Hessian}-unbiased but not FOIU). Strikingly, these are not isolated features of the Neyman-Scott model; SST23 show that the following result holds more generally.

Proposition 7.1 (An impossibility result; SST23, Proposition 6.1) *Among all additive corrections to a general panel data profile loglikelihood that make it FOSU, there does not exist one that also makes it FOIU and FOHU.* □

---

10 This may appear paradoxical at first sight because the CMLE, being semiparametrically efficient (Hahn 1997), is expected to perform better in small samples than its competitors.

11 E.g., in the Neyman-Scott model, the conditional likelihood is not FOHU because it coincides with the IL of SST21, which is not FOHU.

7.10 Relevance for Applied Research

The following observations summarize the key findings of this chapter and may be useful to applied researchers working in this area:

(i) Two pseudolikelihoods can yield identical estimators yet display markedly different behavior when used for LR-based inference. This highlights the fact that identical point estimates do not necessarily imply identical inferential properties.

(ii) When the primary goal is to estimate the common parameters, FOSU pseudolikelihoods should be preferred to non-FOSU ones. Maximizing a FOSU pseudolikelihood typically produces estimators with superior small-sample performance, both in terms of bias and variance, compared to estimators obtained from non-FOSU pseudolikelihoods. Moreover, estimators that correct the bias of the profile-likelihood score not only reduce bias but may also exhibit lower finite-sample variance than the fixed-effects MLE. In settings where n and T are of comparable magnitude, such bias-corrected estimators can therefore serve as viable alternatives to CMLEs, even in models where the latter exist.

(iii) If the focus instead is on LR-based inference, then although FOSU pseudolikelihoods generally outperform their non-FOSU counterparts, the most accurate results arise from pseudolikelihoods that are both FOSU and FOIU. This observation also helps explain why confidence regions based on conditional likelihoods (that is, conditional on sufficient statistics) can achieve excellent coverage even in very short panels.

References

Arellano, Manuel (2003). "Discrete choices with panel data". In: *Investigaciones Econas* XXVII, pp. 423–458.

Arellano, Manuel and Stéphane Bonhomme (2009). "Robust priors in nonlinear panel data models". In: *Econometrica* 77, pp. 489–536.

Arellano, Manuel and Jinyong Hahn (2007). "Understanding bias in nonlinear panel models: Some recent developments". In: *Advances in Economics and Econometrics: Ninth World Congress*. Ed. by R. Blundell, W. Newey, and T. Persson. Vol. 3. Cambridge University Press, Cambridge, UK, pp. 381–409.

Arellano, Manuel and Jinyong Hahn (2016). "A likelihood-based approximate solution to the incidental parameter problem in dynamic nonlinear models with multiple effects". In: *Global Economic Review* 45, pp. 251–274.

Barndorff-Nielsen, O. E. (1983). "On a formula for the distribution of the maximum likelihood estimator". In: *Biometrika* 70, pp. 343–365.

Barndorff-Nielsen, O. E. and Peter Hall (1988). "On the level-error after Bartlett adjustment of the likelihood ratio statistic". In: *Biometrika* 75, pp. 374–378.

Berger, J. O., B. Liseo, and R. Wolpert (1999). "Integrated likelihood functions for eliminating nuisance parameters (with discussion)". In: *Statistical Science* 14, pp. 1–28.

Chamberlain, Gary (1980). "Analysis of covariance with qualitative data". In: *Review of Economic Studies* XLVII, pp. 225–238.

Cox, D. R. and N. Reid (1987). "Parameter orthogonality and approximate conditional inference". In: *Journal of the Royal Statistical Society, Series B* 49, pp. 1–39.

DiCiccio, Thomas J., Michael A. Martin, Steven E. Stern, and G. Alastair Young (1996). "Information bias and adjusted profile likelihoods". In: *Journal of the Royal Statistical Society, Series B* 58, pp. 189–203.

Fernández-Val, Iván (2009). "Fixed effects estimation of structural parameters and marginal effects in panel probit models". In: *Journal of Econometrics* 150, pp. 71–85.

Hahn, Jinyong (1997). "A note on the efficient semiparametric estimation of some exponential panel models". In: *Econometric Theory* 13, pp. 583–588.

Hahn, Jinyong and Whitney K. Newey (2004). "Jackknife and analytical bias reduction for nonlinear panel models". In: *Econometrica* 72, pp. 1295–1319.

Kalbfleisch, J. D. and D. A. Sprott (1970). "Application of likelihood methods to models involving large numbers of parameters (with discussion)". In: *Journal of the Royal Statistical Society, Series B* 32, pp. 175–208.

Kalbfleisch, J. D. and D. A. Sprott (1973). "Marginal and conditional likelihoods". In: *Sankhya, Series A* 35, pp. 311–328.

Lancaster, Tony (2002). "Orthogonal parameters and panel data". In: *Review of Economic Studies* 69, pp. 647–666.

Li, Haihong, Bruce G. Lindsay, and Richard P. Waterman (2003). "Efficiency of projected score methods in rectangular array asymptotics". In: *Journal of the Royal Statistical Society, Series B* 65, pp. 191–208.

McCullagh, Peter and Robert Tibshirani (1990). "A simple method for the adjustment of profile likelihoods". In: *Journal of the Royal Statistical Society, Series B* 52, pp. 325–344.

Mykland, Per A. (1999). "Bartlett identities and large deviations in likelihood theory". In: *Annals of Statistics* 27, pp. 1105–1117.

Neyman, Jerzy and Elizabeth L. Scott (1948). "Consistent estimation from partially consistent observations". In: *Econometrica* 16, pp. 1–32.

Pace, Luigi and Alessandra Salvan (2006). "Adjustments of the profile likelihood from a new perspective". In: *Journal of Statistical Planning and Inference* 136, pp. 3554–3564.

Sartori, Nicola (2003). "Modified profile likelihoods in models with stratum nuisance parameters". In: *Biometrika* 90, pp. 533–549.

Schumann, Martin (2023). "Second-order bias reduction for nonlinear panel data models with fixed effects based on expected quantities". In: *Econometric Theory* 39, pp. 693–736.

Schumann, Martin, Thomas A. Severini, and Gautam Tripathi (2021). "Integrated likelihood based inference for nonlinear panel data models with unobserved effects". In: *Journal of Econometrics* 223, pp. 73–95.

Schumann, Martin, Thomas A. Severini, and Gautam Tripathi (2023). "The role of score and information bias in panel data likelihoods". In: *Journal of Econometrics* 235, pp. 1215–1238.

Severini, Thomas A. (1998). "Likelihood functions for inference in the presence of a nuisance parameter". In: *Biometrika* 85, pp. 507–522.

Severini, Thomas A. (2000). *Likelihood methods in statistics*. Oxford University Press, London, UK.

Severini, Thomas A. (2007). "Integrated likelihood functions for non-Bayesian inference". In: *Biometrika* 94, pp. 529–542.

Chapter 8
Shrinkage Estimators of BLUE for Time Series Regression Models

Abstract The least squares estimator (LSE) seems a natural estimator of linear regression models. Whereas, if the dimension of the vector of regression coefficients is greater than 1 and the residuals are dependent, the best linear unbiased estimator (BLUE), which includes the information of the covariance matrix $\boldsymbol{\Gamma}$ of residual process has a better performance than LSE in the sense of mean square error. As we know the unbiased estimators are generally inadmissible, Senda and Taniguchi (2006) introduced a James-Stein type shrinkage estimator for the regression coefficients based on LSE, where the residual process is a Gaussian stationary process, and provides sufficient conditions such that the James-Stein type shrinkage estimator improves LSE. In this chapter, we propose a shrinkage estimator based on BLUE. Sufficient conditions for this shrinkage estimator to improve BLUE are also given. Furthermore, since $\boldsymbol{\Gamma}$ is infeasible, assuming that $\boldsymbol{\Gamma}$ has a form of $\boldsymbol{\Gamma} = \boldsymbol{\Gamma}(\boldsymbol{\theta})$, we introduce a feasible version of that shrinkage estimator with replacing $\boldsymbol{\Gamma}(\boldsymbol{\theta})$ by $\boldsymbol{\Gamma}(\hat{\boldsymbol{\theta}})$ which is introduced in Toyooka (1986). Additionally, we give the sufficient conditions where the feasible version improves BLUE. Besides, the results of a numerical study confirm our approach. This chapter is mainly based on Xue et al. (2024).

Keywords Shrinkage estimator · LSE · BLUE · Time series regression model · Feasible BLUE · Shrinkage feasible BLUE

8.1 Introduction

It is well known that the unbiased estimators are generally inadmissible. Stein (1956) demonstrated that it is possible to make a uniform improvement with respect to the mean squared error (MSE) when estimating several parameters from the independent normal observations. Furthermore, James and Stein (1961) introduced a particularly simple estimator, of which the improvement was uniform and quite notable near the origin. Numerous James-Stein type estimators have since been proposed for empirical Bayes estimation (Efron and Morris 1975; Lahiri and Rao 1995) and least

M. Taniguchi et al., *Econometrics, Finance, and Time Series Analysis*,
JSS Research Series in Statistics,
https://doi.org/10.1007/978-981-95-8045-3_8

squares estimation (Arnold 1981; Senda and Taniguchi 2006), and so on. Shiraishi et al. (2018) introduced a shrinkage estimator of maximum likelihood estimator (MLE) with a generalized form which includes James-Stein type estimator, and a sufficient condition for the shrinkage estimator to improve the MLE is given. From time series regression models, the best linear unbiased estimator (BLUE) is much better than the least squares estimator (LSE) when the residual process has near unit root (e.g., Taniguchi (1991), pp. 60 and 142). Hence, this study focuses on a multidimensional time series regression estimation using BLUE.

Suppose that $\{\boldsymbol{y}(1), \ldots, \boldsymbol{y}(n)\}$ is a sequence of p-dimensional i.i.d. random vectors distributed as $N_p(\boldsymbol{\theta}, \boldsymbol{I}_p)$, where $\boldsymbol{I}_p$ is a $p \times p$ identity matrix. The sample mean $\bar{\boldsymbol{Y}}_n \equiv n^{-1} \sum_{t=1}^{n} \boldsymbol{y}(t)$ seems the most fundamental and natural estimator of $\boldsymbol{\theta}$. However, if $p \geq 3$, Stein (1956) demonstrated that $\bar{\boldsymbol{Y}}_n$ is not admissible with respect to the mean squared error (MSE) loss function.

Furthermore, James and Stein (1961) proposed the following shrinkage estimator:

$$\hat{\boldsymbol{\theta}}_n = \left(1 - \frac{p-2}{n\|\bar{\boldsymbol{Y}}_n\|^2}\right) \bar{\boldsymbol{Y}}_n,$$

which improves $\bar{\boldsymbol{Y}}_n$ with respect to MSE when $p \geq 3$.

In the case when $\boldsymbol{y}(t)$ is dependent, that is, a p-dimensional Gaussian process, Taniguchi and Hirukawa (2005) evaluated the difference between MSE of $\bar{\boldsymbol{Y}}_n$ and MSE of $\hat{\boldsymbol{\theta}}_n$, that is,

$$\text{DMSE} \equiv \text{MSE}(\bar{\boldsymbol{Y}}_n) - \text{MSE}(\hat{\boldsymbol{\theta}}_n)$$

up to $O(n^{-1})$, which is referred to as the third-order asymptotics. They then provided a sufficient condition so that $\hat{\boldsymbol{\theta}}_n$ improves $\bar{\boldsymbol{Y}}_n$ up to third order. For time series data, estimating the trend is very important in various fields such as engineering, natural sciences, econometrics, and so on. Usually, we estimate the coefficients of regression part by LSE. For a p-dimensional time series regression model,

$$\boldsymbol{y}(t) = \boldsymbol{B}^\top \boldsymbol{x}(t) + \boldsymbol{\varepsilon}(t),$$

where $\{\boldsymbol{\varepsilon}(t)\}$ is a p-dimensional Gaussian process and $\boldsymbol{x}(t)$s are q-dimensional regressors. Senda and Taniguchi (2006) introduced a shrinkage estimator for the vectorized $\boldsymbol{\beta} = \text{vec}\{\boldsymbol{B}\}$:

$$\hat{\boldsymbol{\beta}}_{\text{JS}} \equiv \{1 - (\text{function of standardized } \hat{\boldsymbol{\beta}}_{\text{LS}})\}\, \hat{\boldsymbol{\beta}}_{\text{LS}},$$

where $\hat{\boldsymbol{\beta}}_{\text{LS}}$ denotes the LSE of $\boldsymbol{\beta}$. They then provided a sufficient condition such that $\hat{\boldsymbol{\beta}}_{\text{JS}}$ improves $\hat{\boldsymbol{\beta}}_{\text{LS}}$ with respect to MSE. By contrast, if the dimension of the vector of regression coefficients is greater than 1 and the residuals are dependent, the BLUE that includes the information of the covariance matrix $\boldsymbol{\Gamma}$ of residual process has a better performance than LSE in the sense of mean square error (see also Taniguchi (1991), pp. 60 and 142). Using the BLUE $\hat{\boldsymbol{\beta}}_{\text{BLUE}}$ of $\boldsymbol{\beta}$, instead of $\hat{\boldsymbol{\beta}}_{\text{LS}}$, we propose a shrinkage estimator:

$$\hat{\boldsymbol{\beta}} \equiv \{1 - (\text{function of standardized } \hat{\boldsymbol{\beta}}_{\text{BLUE}})\}\hat{\boldsymbol{\beta}}_{\text{BLUE}}. \tag{8.1}$$

We then provide a sufficient condition such that $\hat{\boldsymbol{\beta}}$ improves $\hat{\boldsymbol{\beta}}_{\text{BLUE}}$. Although $\hat{\boldsymbol{\beta}}_{\text{BLUE}}$ is the best linear estimator of $\boldsymbol{\beta}$, it uses the information from the covariance matrix $\boldsymbol{\Gamma}$ of the residual process. Usually, it has the form of $\boldsymbol{\Gamma} = \boldsymbol{\Gamma}(\theta)$, where θ is an unknown parameter of $\{\boldsymbol{\varepsilon}(t)\}$. If θ is estimated by $\hat{\theta}$, we use $\boldsymbol{\Gamma}(\hat{\theta})$ to substitute $\boldsymbol{\Gamma}(\theta)$ in the definition of $\hat{\boldsymbol{\beta}}_{\text{BLUE}}$, denoting the new estimator by $\tilde{\boldsymbol{\beta}}_{\text{BLUE}}$. Toyooka (1986) demonstrated that when θ is one-dimensional, and the residual process is univariate, $\text{MSE}(\hat{\boldsymbol{\beta}}_{\text{BLUE}}) - \text{MSE}(\tilde{\boldsymbol{\beta}}_{\text{BLUE}}) = o(n^{-2})$, which implies that $\tilde{\boldsymbol{\beta}}_{\text{BLUE}}$ is asymptotically equivalent to $\hat{\boldsymbol{\beta}}_{\text{BLUE}}$ under natural conditions. As we know that when the dimension of θ is greater than 1, and the residual process is multivariate, from Chapter 6 of Taniguchi and Kakizawa (2020), under certain conditions we can also construct estimators of θ with similar properties as in Toyooka (1986). Hence, we introduce a feasible approximator of $\hat{\boldsymbol{\beta}}$ by replacing $\hat{\boldsymbol{\beta}}_{\text{BLUE}}$ by $\tilde{\boldsymbol{\beta}}_{\text{BLUE}}$ in (8.1), and also present a sufficient condition where the new estimator improves $\hat{\boldsymbol{\beta}}_{\text{BLUE}}$ at third order.

The remainder of this chapter is organized as follows. Section 8.2 describes the elements of time series regression models, LSE and BLUE. Section 8.3 presents the asymptotic theory of $\hat{\boldsymbol{\beta}}_{\text{BLUE}}$. We compare the difference between $\hat{\boldsymbol{\beta}}$ and $\hat{\boldsymbol{\beta}}_{\text{BLUE}}$ in two situations and provide a sufficient condition such that $\hat{\boldsymbol{\beta}}$ improves $\hat{\boldsymbol{\beta}}_{\text{BLUE}}$. In Sect. 8.4, a feasible approximator of $\hat{\boldsymbol{\beta}}$ is introduced, and the theoretical support for its improvement on MSE is clarified. Section 8.5 provides numerical studies on MSEs of those estimators mentioned. The results confirm the theorems well.

8.2 Elements of Time Series Regression Models

Suppose that $\boldsymbol{y}(t) = (y_1(t), \ldots, y_p(t))^\top$ is generated by

$$\boldsymbol{y}(t) = \boldsymbol{B}^\top \boldsymbol{x}(t) + \boldsymbol{\varepsilon}(t), \qquad t \in \{1, \ldots, n\}, \tag{8.2}$$

where $\boldsymbol{B} = (b_{ij})$ is a $q \times p$ matrix of unknown coefficients, and $\{\boldsymbol{\varepsilon}(t)\}$ is a p-dimensional Gaussian stationary process with mean $\boldsymbol{0}$ and spectral density matrix $\boldsymbol{f}(\lambda)$. Here, $\boldsymbol{x}(t) = (x_1(t), \ldots, x_q(t))^\top$ is a $q \times 1$ vector of regressors satisfying the assumptions below. Let $d_i^2(n) = \sum_{t=1}^{n} \{x_i(t)\}^2$, $i \in \{1, \ldots, q\}$, and $d_i(n)$ is the positive square root of $d_i^2(n)$.

Assumption 8.1 $\lim_{n \to \infty} d_i(n)^2 = \infty, i \in \{1, \ldots, q\}$. □

Assumption 8.2 $\lim_{n \to \infty} \{x_i(n)\}^2 / d_i(n)^2 = 0, i \in \{1, \ldots, q\}$, and $x_i(n)/n^{\kappa_i} = O(1)$ for some $\kappa_i \geq 0, i \in \{1, \ldots, q\}$.
Set $\min\{\kappa_1, \ldots, \kappa_q\} = \kappa$. □

Assumption 8.3 For every $i, j \in \{1, \ldots, q\}$, there exists the limit

$$\rho_{ij}(h) \equiv \lim_{n\to\infty} \frac{\sum_{t=1}^{n} x_i(t)x_j(t+h)}{d_i(n)d_j(n)}.$$

Let $\boldsymbol{R}(h) \equiv (\rho_{ij}(h))$. □

Assumption 8.4 $\boldsymbol{R}(0)$ is nonsingular.

Then we can write

$$\boldsymbol{R}(h) = \int_{-\pi}^{\pi} \mathrm{e}^{ih\lambda} \boldsymbol{M}(d\lambda),$$

where $\boldsymbol{M}(\lambda)$ is a matrix function whose increments are Hermitian non-negative. □

Assumption 8.5 $\{\boldsymbol{\varepsilon}(t)\}$ has a piecewise continuous and positive-definite spectral density matrix $\boldsymbol{f}(\lambda)$ with no discontinuities at the jumps of $\boldsymbol{M}(\lambda)$. □

We write (8.2) in the tensor form:

$$\boldsymbol{Y} = (\boldsymbol{I}_p \otimes \boldsymbol{X})\boldsymbol{\beta} + \boldsymbol{u} = \boldsymbol{U}\boldsymbol{\beta} + \boldsymbol{u}$$

where $y_i(t)$ is in row $(i-1)n+t$ of $\boldsymbol{Y}$, $\varepsilon_i(t)$ is in row $(i-1)n+t$ of $\boldsymbol{u}$, which implies that $\boldsymbol{u}$ is the tensor form of $\boldsymbol{\varepsilon}$, $\boldsymbol{X}$ has $x_j(t)$ in row t column j, $\boldsymbol{U} \equiv \boldsymbol{I}_p \otimes \boldsymbol{X}$, and $\boldsymbol{\beta}$ has b_{ij} in row $(j-1)q+i$. As an estimator of $\boldsymbol{\beta}$, one of the most fundamental candidate is the LSE $\hat{\boldsymbol{\beta}}_{\mathrm{LS}} \equiv (\boldsymbol{U}^\top \boldsymbol{U})^{-1}\boldsymbol{U}^\top \boldsymbol{Y}$. When $\boldsymbol{\varepsilon}(t)$s are i.i.d., we know that $\hat{\boldsymbol{\beta}}_{\mathrm{LS}}$ is not admissible if $pq \geq 3$.

Senda and Taniguchi (2006) proposed an alternative estimator

$$\hat{\boldsymbol{\beta}}_{\mathrm{JS}} \equiv \left(1 - \frac{c}{\|(\boldsymbol{I}_p \otimes \boldsymbol{D}_n)\hat{\boldsymbol{\beta}}_{\mathrm{LS}}\|^2}\right)(\hat{\boldsymbol{\beta}}_{\mathrm{LS}} - \boldsymbol{b}) + \boldsymbol{b}, \tag{8.3}$$

known as the James-Stein estimator for $\boldsymbol{\beta}$. Here $\boldsymbol{D}_n = \mathrm{diag}\{d_1(n), \ldots .d_q(n)\}$, and $c > 0$. Moreover, Senda and Taniguchi (2006) documented that under certain conditions, the risk of $\hat{\boldsymbol{\beta}}_{\mathrm{JS}}$ is smaller than that of $\hat{\boldsymbol{\beta}}_{\mathrm{LS}}$ under the MSE loss function:

$$\mathrm{MSE}_n(\boldsymbol{\phi}(\boldsymbol{U}, \boldsymbol{Y})) \equiv \mathrm{E}\left[\|(\boldsymbol{I}_p \otimes \boldsymbol{D}_n)(\boldsymbol{\phi}(\boldsymbol{U}, \boldsymbol{Y}) - \boldsymbol{\beta})\|^2\right],$$

where $\boldsymbol{\phi}(\boldsymbol{U}, \boldsymbol{Y})$ is an arbitrary estimator of $\boldsymbol{\beta}$.

Instead of $\hat{\boldsymbol{\beta}}_{\mathrm{LS}}$ in (8.3) with $\boldsymbol{b} = \boldsymbol{0}$, without loss of generality, we propose a shrinkage estimator $\hat{\boldsymbol{\beta}}$ based on the BLUE estimator of $\boldsymbol{\beta}$:

$$\hat{\boldsymbol{\beta}} \equiv \left(1 - \frac{c}{\|(\boldsymbol{I}_p \otimes \boldsymbol{D}_n)\hat{\boldsymbol{\beta}}_{\mathrm{BLUE}}\|^2}\right)\hat{\boldsymbol{\beta}}_{\mathrm{BLUE}}, \tag{8.4}$$

where

$$\hat{\boldsymbol{\beta}}_{\text{BLUE}} \equiv (\boldsymbol{U}^\top \boldsymbol{\Gamma}^{-1} \boldsymbol{U})^{-1} \boldsymbol{U}^\top \boldsymbol{\Gamma}^{-1} \boldsymbol{Y},$$

and $\boldsymbol{\Gamma}$ is the autocovariance matrix of $\boldsymbol{u}$. In what follows, we deal with the case where $\left(1 - c/\|(\boldsymbol{I}_p \otimes \boldsymbol{D}_n)\hat{\boldsymbol{\beta}}_{\text{BLUE}}\|^2\right) > 0$ a.s. If we choose c and n suitably, we know $\left(1 - c/\|(\boldsymbol{I}_p \otimes \boldsymbol{D}_n)\hat{\boldsymbol{\beta}}_{\text{BLUE}}\|^2\right) > 0$ a.s. (see Remark 8.1).

8.3 Asymptotic Theory for BLUE

For the James-Stein estimator based on LSE, $\hat{\boldsymbol{\beta}}_{\text{JS}}$, Senda and Taniguchi (2006) illustrated the conditions under which the estimator improves LSE by evaluating the difference between MSEs of $\hat{\boldsymbol{\beta}}_{\text{JS}}$ and $\hat{\boldsymbol{\beta}}_{\text{LS}}$.

In this section, we evaluate

$$\begin{aligned} \text{DMSE}_n(\hat{\boldsymbol{\beta}}_{\text{BLUE}}, \hat{\boldsymbol{\beta}}) &\equiv \text{e}\{\|(\boldsymbol{I}_p \otimes \boldsymbol{D}_n)(\hat{\boldsymbol{\beta}}_{\text{BLUE}} - \boldsymbol{\beta})\|^2\} - \text{e}\{\|(\boldsymbol{I}_p \otimes \boldsymbol{D}_n)(\hat{\boldsymbol{\beta}} - \boldsymbol{\beta})\|^2\} \\ &= -c^2 \text{e}\left[\frac{1}{\|(\boldsymbol{I}_p \otimes \boldsymbol{D}_n)\hat{\boldsymbol{\beta}}_{\text{BLUE}}\|^2}\right] \\ &\quad + 2c\left(1 - \text{e}\left[\frac{< (\boldsymbol{I}_p \otimes \boldsymbol{D}_n)\boldsymbol{\beta}, (\boldsymbol{I}_p \otimes \boldsymbol{D}_n)\hat{\boldsymbol{\beta}}_{\text{BLUE}} >}{\|(\boldsymbol{I}_p \otimes \boldsymbol{D}_n)\hat{\boldsymbol{\beta}}_{\text{BLUE}}\|^2}\right]\right), \end{aligned} \tag{8.5}$$

when $pq \geq 3$. Note that $\text{Cov}\{(\boldsymbol{I}_p \otimes \boldsymbol{D}_n)(\hat{\boldsymbol{\beta}}_{\text{BLUE}} - \boldsymbol{\beta})\} = \boldsymbol{C}_n$, and from Theorem 5 in Chapter 7 in Hannan (1970) or Chapter 7 in Grenander and Rosenblatt (1957), $\boldsymbol{C} \equiv \lim_{n\to\infty} \boldsymbol{C}_n$, where $\boldsymbol{C}_n = (\boldsymbol{I}_p \otimes \boldsymbol{D}_n)(\boldsymbol{U}^\top \boldsymbol{\Gamma}^{-1} \boldsymbol{U})^{-1}(\boldsymbol{I}_p \otimes \boldsymbol{D}_n)$, and $\boldsymbol{C} = \left(\int (2\pi \boldsymbol{f}(\lambda))^{-1} \otimes \text{d}\boldsymbol{M}(\lambda)\right)^{-1}$.

Theorem 8.1 *In the case where $\boldsymbol{\beta} = \boldsymbol{0}$, suppose that Assumptions 8.1–8.5 hold and that $pq \geq 3$. Then,*

(i)

$$\begin{aligned} c\left\{2 - \frac{c}{pq-2}\left(\frac{\nu_{pq,n}}{\nu_{1,n}}\right)^{pq/2} \frac{1}{\nu_{pq,n}}\right\} &\leq \text{DMSE}_n(\hat{\boldsymbol{\beta}}_{\text{BLUE}}, \hat{\boldsymbol{\beta}}) \\ &\leq c\left\{2 - \frac{c}{pq-2}\left(\frac{\nu_{1,n}}{\nu_{pq,n}}\right)^{pq/2} \frac{1}{\nu_{1,n}}\right\}, \end{aligned} \tag{8.6}$$

which implies that $\hat{\boldsymbol{\beta}}$ improves $\hat{\boldsymbol{\beta}}_{\text{BLUE}}$ if the left-hand side (LHS) of (8.6) is positive;

(ii)

$$c\left\{2-\frac{c}{pq-2}\left(\frac{\nu_{pq}}{\nu_1}\right)^{pq/2}\frac{1}{\nu_{pq}}\right\}\le\lim_{n\to\infty}\mathrm{DMSE}_n(\hat{\boldsymbol{\beta}}_{\mathrm{BLUE}},\hat{\boldsymbol{\beta}})$$
$$\le c\left\{2-\frac{c}{pq-2}\left(\frac{\nu_1}{\nu_{pq}}\right)^{pq/2}\frac{1}{\nu_1}\right\},\quad(8.7)$$

which implies that $\hat{\boldsymbol{\beta}}$ improves $\hat{\boldsymbol{\beta}}_{\mathrm{BLUE}}$ asymptotically if the LHS of (8.7) is positive, where $\nu_{1,n},\ldots,\nu_{pq,n}$ ($\nu_{1,n}\le\cdots\le\nu_{pq,n}$) and $\nu_1,\ldots,\nu_{pq}$ ($\nu_1\le\cdots\le\nu_{pq}$) are the eigenvalues of $\boldsymbol{C}_n$ and $\boldsymbol{C}$, respectively.

Proof. In the case of $\boldsymbol{\beta}=\boldsymbol{0}$, from (8.5), we have

$$\mathrm{DMSE}_n(\hat{\boldsymbol{\beta}}_{\mathrm{BLUE}},\hat{\boldsymbol{\beta}})=-c^2\mathrm{e}\left[\frac{1}{\|(\boldsymbol{I}_p\otimes\boldsymbol{D}_n)\hat{\boldsymbol{\beta}}_{\mathrm{BLUE}}\|^2}\right]+2c.$$

Noticing that $\{\boldsymbol{\varepsilon}(t)\}$ is a Gaussian process, and that

$$(\boldsymbol{I}_p\otimes\boldsymbol{D}_n)\hat{\boldsymbol{\beta}}_{\mathrm{BLUE}}=(\boldsymbol{I}_p\otimes\boldsymbol{D}_n)(\hat{\boldsymbol{\beta}}_{\mathrm{BLUE}}-\boldsymbol{\beta})=(\boldsymbol{I}_p\otimes\boldsymbol{D}_n)(\boldsymbol{U}^\top\boldsymbol{\Gamma}^{-1}\boldsymbol{U})^{-1}\boldsymbol{U}^\top\boldsymbol{\Gamma}^{-1}\boldsymbol{u},$$

we have

$$(\boldsymbol{I}_p\otimes\boldsymbol{D}_n)\hat{\boldsymbol{\beta}}_{\mathrm{BLUE}}\sim N(\boldsymbol{0},\boldsymbol{C}_n),$$

and because C_n is a symmetric matrix, a $pq\times pq$ orthogonal matrix P exists that diagonalizes $\boldsymbol{C}_n$, such that

$$P^\top(\boldsymbol{I}_p\otimes\boldsymbol{D}_n)\hat{\boldsymbol{\beta}}_{\mathrm{BLUE}}\sim N(\boldsymbol{0},\mathrm{diag}\{\nu_{1,n},\ldots,\nu_{pq,n}\}).$$

By following the proof line of Theorem 1 in Taniguchi and Hirukawa (2005), (8.6) holds. Furthermore, by taking the limit of each term in (8.6), (8.7) holds. □

Remark 8.1

(i) If $\boldsymbol{\beta}=\boldsymbol{0}$, from Theorem 8.1, c can be chosen from $(0,2(pq-2)\nu_{1,n}^{pq/2}/\nu_{pq,n}^{pq/2-1})$, which means that
$0<\mathrm{e}[c/\|(\boldsymbol{I}_p\otimes\boldsymbol{D}_n)\hat{\boldsymbol{\beta}}_{BLUE}\|^2]<2$. Hence, for any $\varepsilon>0$, if we take a sufficiently small $c_0>0$ and so that $\mathrm{e}[c_0/\|(\boldsymbol{I}_p\otimes\boldsymbol{D}_n)\hat{\boldsymbol{\beta}}_{BLUE}\|^2]<\varepsilon$, we can see $\left(1-c_0/\|(\boldsymbol{I}_p\otimes\boldsymbol{D}_n)\hat{\boldsymbol{\beta}}_{\mathrm{BLUE}}\|^2\right)>0$ a.s. by applying Markov's inequality.

(ii) For $\boldsymbol{\beta}\ne\boldsymbol{0}$, we notice that as sample size increases, $\left(1-c/\|(\boldsymbol{I}_p\otimes\boldsymbol{D}_n)\hat{\boldsymbol{\beta}}_{\mathrm{BLUE}}\|^2\right)$ goes to 1 which implies the positivity.

Next, we consider the case in which $\boldsymbol{\beta}\ne\boldsymbol{0}$.

Theorem 8.2 *In the case of* $\boldsymbol{\beta} \neq \mathbf{0}$, *suppose that Assumptions 8.1–8.5 hold and that* $p \geq 3$. *Then, if* $c > 0$,

(i)

$$\text{DMSE}_n(\hat{\boldsymbol{\beta}}_{\text{BLUE}}, \hat{\boldsymbol{\beta}}) = \frac{2c}{\|(\boldsymbol{I}_p \otimes \boldsymbol{D}_n)\boldsymbol{\beta}\|^2}\{\Delta_n(c, \boldsymbol{\beta}) + o(1)\}, \tag{8.8}$$

where

$$\Delta_n(c, \boldsymbol{\beta}) = \text{tr}\boldsymbol{C}_n - \frac{2\boldsymbol{\beta}^\top(\boldsymbol{I}_p \otimes \boldsymbol{D}_n)\boldsymbol{C}_n(\boldsymbol{I}_p \otimes \boldsymbol{D}_n)\boldsymbol{\beta}}{\|(\boldsymbol{I}_p \otimes \boldsymbol{D}_n)\boldsymbol{\beta}\|^2} - \frac{c}{2}.$$

Here, we have the inequality

$$\Delta_n(c, \boldsymbol{\beta}) \geq \sum_{j=1}^{pq-1} \nu_{j,n} - \nu_{pq,n} - \frac{c}{2}, \tag{8.9}$$

which implies that $\hat{\boldsymbol{\beta}}$ *improves* $\hat{\boldsymbol{\beta}}_{\text{BLUE}}$ *up to* $n^{-2\kappa-1}$*-order if the right-hand side (RHS) of (8.9) is positive, where* κ *is same as mentioned in Assumption 8.2.*

(ii) *Taking the limit of (8.9), we have*

$$\lim_{n\to\infty} \Delta_n(c, \boldsymbol{\beta}) \geq \sum_{j=1}^{pq-1} \nu_j - \nu_{pq} - \frac{c}{2}, \tag{8.10}$$

which implies that $\hat{\boldsymbol{\beta}}$ *improves* $\hat{\boldsymbol{\beta}}_{\text{BLUE}}$ *asymptotically up to* $n^{-2\kappa-1}$*-order if the RHS of (8.10) is positive.*

Proof.

$$\begin{aligned}\text{DMSE}_n(\hat{\boldsymbol{\beta}}_{\text{BLUE}}, \hat{\boldsymbol{\beta}}) &= -c^2\text{e}\left[\frac{1}{\|(\boldsymbol{I}_p \otimes \boldsymbol{D}_n)\hat{\boldsymbol{\beta}}_{\text{BLUE}}\|^2}\right] \\ &\quad + 2c\text{e}\left[< (\boldsymbol{I}_p \otimes \boldsymbol{D}_n)(\hat{\boldsymbol{\beta}}_{\text{BLUE}} - \boldsymbol{\beta}), \frac{(\boldsymbol{I}_p \otimes \boldsymbol{D}_n)\hat{\boldsymbol{\beta}}_{\text{BLUE}}}{\|(\boldsymbol{I}_p \otimes \boldsymbol{D}_n)\hat{\boldsymbol{\beta}}_{\text{BLUE}}\|^2} >\right] \\ &= -c^2\text{e}_1 + 2c\text{e}_2 \quad \text{(say)}.\end{aligned}$$

As in Senda and Taniguchi (2006), we only need to show that

$$\frac{\|(\boldsymbol{I}_p \otimes \boldsymbol{D}_n)(\hat{\boldsymbol{\beta}}_{\text{BLUE}} - \boldsymbol{\beta})\|}{\|(\boldsymbol{I}_p \otimes \boldsymbol{D}_n)\boldsymbol{\beta}\|} = o(1) \quad \text{a.s..} \tag{8.11}$$

Without loss of generality we assume that $p = 1$. Then

$$(\boldsymbol{I}_p \otimes \boldsymbol{D}_n)(\hat{\boldsymbol{\beta}}_{\text{BLUE}} - \boldsymbol{\beta}) = \boldsymbol{D}_n(\hat{\boldsymbol{\beta}}_{\text{BLUE}} - \boldsymbol{\beta}).$$

Note that

$$\boldsymbol{D}_n(\hat{\boldsymbol{\beta}}_{\text{BLUE}} - \boldsymbol{\beta}) = \boldsymbol{D}_n(\boldsymbol{X}^\top \boldsymbol{\Gamma}^{-1}\boldsymbol{X})^{-1}\boldsymbol{X}^\top \boldsymbol{\Gamma}^{-1}\boldsymbol{\varepsilon} = (\boldsymbol{D}_n^{-1}\boldsymbol{X}^\top \boldsymbol{\Gamma}^{-1}\boldsymbol{X}\boldsymbol{D}_n^{-1})^{-1}\boldsymbol{D}_n^{-1}\boldsymbol{X}^\top \boldsymbol{\Gamma}^{-1}\boldsymbol{\varepsilon}$$

and that

$$(\boldsymbol{D}_n^{-1}\boldsymbol{X}^\top \boldsymbol{\Gamma}^{-1}\boldsymbol{X}\boldsymbol{D}_n^{-1})^{-1} = \boldsymbol{O}(1)$$

from Assumption 8.3. Then we evaluate $\boldsymbol{X}^\top \boldsymbol{\Gamma}^{-1}\boldsymbol{\varepsilon}$. Noticing that the elements of $\boldsymbol{\Gamma}^{-1}\boldsymbol{\varepsilon}$ constitute a Gaussian process with spectral density $f^{-1}(\lambda)$ as $n \to \infty$, by Theorem 3.2 of He (1995), a constant $m > 0$ exists such that for $i \in \{1, \ldots, q\}$,

$$\limsup_{n\to\infty} \frac{1}{\sqrt{n^{2\kappa_i+1}\log n}} \left|(1^{\kappa_i}, \ldots, n^{\kappa_i})\boldsymbol{\Gamma}^{-1}\boldsymbol{\varepsilon}\right| \le m, \quad \text{a.s.,}$$

which, together with Assumption 8.2, shows that for each i and some constant $m' > 0$,

$$|(x_i(1), \ldots, x_i(n))\boldsymbol{\Gamma}^{-1}\boldsymbol{\varepsilon}| \le m'|(1^{\kappa_i}, \ldots, n^{\kappa_i})\boldsymbol{\Gamma}^{-1}\boldsymbol{\varepsilon}| = O(\sqrt{n^{2\kappa_i+1}\log n}), \quad \text{a.s..} \quad (8.12)$$

From (8.12), it shows that

$$\left|\frac{1}{d_i(n)}(x_i(1), \ldots, x_i(n))\boldsymbol{\Gamma}^{-1}\boldsymbol{\varepsilon}\right| = O\left(\frac{\sqrt{n^{2\kappa_i+1}\log n}}{d_i(n)}\right) = O(\sqrt{\log n}).$$

Hence,

$$\frac{\|(\boldsymbol{I}_p \otimes \boldsymbol{D}_n)(\hat{\boldsymbol{\beta}}_{\text{BLUE}} - \boldsymbol{\beta})\|}{\|(\boldsymbol{I}_p \otimes \boldsymbol{D}_n)\boldsymbol{\beta}\|} = o(1),$$

that is, (8.11) holds. From (8.11) it seems that

$$\begin{aligned}
\frac{1}{\|(\boldsymbol{I}_p \otimes \boldsymbol{D}_n)\hat{\boldsymbol{\beta}}_{\text{BLUE}}\|^2} &= \frac{1}{\|(\boldsymbol{I}_p \otimes \boldsymbol{D}_n)(\hat{\boldsymbol{\beta}}_{\text{BLUE}} - \boldsymbol{\beta}) + (\boldsymbol{I}_p \otimes \boldsymbol{D}_n)\boldsymbol{\beta}\|^2} \\
&= \frac{1}{\|(\boldsymbol{I}_p \otimes \boldsymbol{D}_n)\boldsymbol{\beta}\|^2} \\
&\quad \left\{1 - \frac{2 < (\boldsymbol{I}_p \otimes \boldsymbol{D}_n)\boldsymbol{\beta}, (\boldsymbol{I}_p \otimes \boldsymbol{D}_n)(\hat{\boldsymbol{\beta}}_{\text{BLUE}} - \boldsymbol{\beta}) >}{\|(\boldsymbol{I}_p \otimes \boldsymbol{D}_n)\boldsymbol{\beta}\|^2} + o(1)\right\}, \quad \text{a.s.}
\end{aligned}$$

Therefore, using Fatou's Lemma,

$$\begin{aligned}
\mathrm{e}_1 &= \frac{1}{\|(\boldsymbol{I}_p \otimes \boldsymbol{D}_n)\boldsymbol{\beta}\|^2}[1 + o(1)], \\
\mathrm{e}_2 &= \frac{1}{\|(\boldsymbol{I}_p \otimes \boldsymbol{D}_n)\boldsymbol{\beta}\|^2} \times \left\{-2\frac{\boldsymbol{\beta}^\top(\boldsymbol{I}_p \otimes \boldsymbol{D}_n)\boldsymbol{C}_n(\boldsymbol{I}_p \otimes \boldsymbol{D}_n)\boldsymbol{\beta}}{\|(\boldsymbol{I}_p \otimes \boldsymbol{D}_n)\boldsymbol{\beta}\|^2} + \mathrm{tr}\boldsymbol{C}_n + o(1)\right\}.
\end{aligned}$$

Thus,

$$\begin{aligned}
\text{DMSE}_n(\hat{\boldsymbol{\beta}}_{\text{BLUE}}, \hat{\boldsymbol{\beta}}) &= -c^2 e_1 + 2ce_2 \\
&= \frac{1}{\|(\boldsymbol{I}_p \otimes \boldsymbol{D}_n)\boldsymbol{\beta}\|^2} \left[-c^2[1 + o(1)] \right. \\
&\quad \left. +2c \left\{ -2\frac{\boldsymbol{\beta}^\top (\boldsymbol{I}_p \otimes \boldsymbol{D}_n)\boldsymbol{C}_n(\boldsymbol{I}_p \otimes \boldsymbol{D}_n)\boldsymbol{\beta}}{\|(\boldsymbol{I}_p \otimes \boldsymbol{D}_n)\boldsymbol{\beta}\|^2} + \text{tr}\boldsymbol{C}_n + o(1) \right\} \right] \\
&= \frac{2c}{\|(\boldsymbol{I}_p \otimes \boldsymbol{D}_n)\boldsymbol{\beta}\|^2} \left\{ \Delta_n(c, \boldsymbol{\beta}) + o(1) \right\}.
\end{aligned}$$

Hence, when $c > 0$ and $\Delta_n(c, \boldsymbol{\beta}) + o(1) > 0$, $\hat{\boldsymbol{\beta}}$ improves $\hat{\boldsymbol{\beta}}_{\text{BLUE}}$. Noticing that $\text{tr}\boldsymbol{C}_n = \sum_{j=1}^{pq} \nu_{j,n}$, and that for an arbitrary vector $\boldsymbol{v}$, $\nu_{pq,n} \geq \boldsymbol{v}^\top \boldsymbol{C}_n \boldsymbol{v}/\|\boldsymbol{v}\|^2$, we have $\Delta_n(c, \boldsymbol{\beta}) \geq \sum_{j=1}^{pq-1} \nu_{j,n} - \nu_{pq,n} - c/2$, which leads to the conclusion. From Assumption 8.2, as $\kappa = \min\{\kappa_1, \ldots, \kappa_q\}$, we have $\|(\boldsymbol{I}_p \otimes \boldsymbol{D}_n)\boldsymbol{\beta}\|^2 = \Omega(n^{2\kappa+1})$. Hence $\hat{\boldsymbol{\beta}}$ improves $\hat{\boldsymbol{\beta}}_{\text{BLUE}}$ up to $n^{-2\kappa-1}$-order. □

8.4 A Feasible Approximator of $\hat{\boldsymbol{\beta}}$

Noticing that $\hat{\boldsymbol{\beta}}_{\text{BLUE}}$ contains the unknown covariance matrix $\boldsymbol{\Gamma}$, which means that $\hat{\boldsymbol{\beta}}_{\text{BLUE}}$ is an infeasible estimator, in what follows, we consider the situation where the spectral density matrix $\boldsymbol{f}(\lambda)$ belongs to a fitted model set $\{\boldsymbol{f}_{\boldsymbol{\theta}}(\lambda); \boldsymbol{\theta} \in \Theta\}$. Thus $\boldsymbol{\Gamma}$ is a functional matrix of $\boldsymbol{\theta}$, that is, $\boldsymbol{\Gamma}(\boldsymbol{\theta})$. By estimating $\boldsymbol{\theta}$, we introduce $\tilde{\boldsymbol{\beta}}$ as a feasible approximator of $\hat{\boldsymbol{\beta}}$, defined as

$$\tilde{\boldsymbol{\beta}} \equiv \left(1 - \frac{c}{\|(\boldsymbol{I}_p \otimes \boldsymbol{D}_n)\tilde{\boldsymbol{\beta}}_{\text{BLUE}}\|^2} \right) \tilde{\boldsymbol{\beta}}_{\text{BLUE}},$$

where

$$\tilde{\boldsymbol{\beta}}_{\text{BLUE}} \equiv (\boldsymbol{U}^\top \boldsymbol{\Gamma}^{-1}(\hat{\boldsymbol{\theta}})\boldsymbol{U})^{-1} \boldsymbol{U}^\top \boldsymbol{\Gamma}^{-1}(\hat{\boldsymbol{\theta}})\boldsymbol{Y},$$

and $\hat{\boldsymbol{\theta}}$ is an estimator of θ using $\hat{\boldsymbol{\beta}}_{\text{LS}}$. Let the estimated residual of $\boldsymbol{u}$ be

$$\tilde{\boldsymbol{u}} = \boldsymbol{Y} - \boldsymbol{U}\hat{\boldsymbol{\beta}}_{\text{LS}} = \boldsymbol{Y} - \boldsymbol{U}(\boldsymbol{U}^\top \boldsymbol{U})^{-1}\boldsymbol{U}^\top \boldsymbol{Y}.$$

Noticing that $\tilde{\boldsymbol{u}} = \left\{ \boldsymbol{I}_{pn} - \boldsymbol{U}(\boldsymbol{U}^\top \boldsymbol{U})^{-1}\boldsymbol{U}^\top \right\} \boldsymbol{u}$, which is vectorized from $\left\{ \boldsymbol{I}_n - \boldsymbol{X}(\boldsymbol{X}^\top \boldsymbol{X})^{-1}\boldsymbol{X}^\top \right\} \boldsymbol{\varepsilon}$, we define $\left\{ \boldsymbol{I}_n - \boldsymbol{X}(\boldsymbol{X}^\top \boldsymbol{X})^{-1}\boldsymbol{X}^\top \right\} \boldsymbol{\varepsilon}$ as an estimator of $\boldsymbol{\varepsilon}$, and denote it by $\tilde{\boldsymbol{\varepsilon}}$, where $\boldsymbol{\varepsilon} = \{\boldsymbol{\varepsilon}(t)^\top\}^\top$. Let $\hat{\boldsymbol{\theta}}$ be defined by

$$\hat{\boldsymbol{\theta}} = \arg\min_{\boldsymbol{\theta} \in \Theta} D(\boldsymbol{I}_{\tilde{\boldsymbol{\varepsilon}}}, \boldsymbol{\theta}),$$

where $D(\cdot,\cdot)$ is a quasi-Gaussian maximum likelihood functional for $\tilde{\boldsymbol{\varepsilon}}$:

$$D(\boldsymbol{I}_{\tilde{\boldsymbol{\varepsilon}}}, \boldsymbol{\theta}) = \int_{-\pi}^{\pi} [\log \det \boldsymbol{f}_{\boldsymbol{\theta}} + \mathrm{tr}\{\boldsymbol{f}_{\boldsymbol{\theta}}^{-1} \boldsymbol{I}_{\tilde{\boldsymbol{\varepsilon}}}\}]\, \mathrm{d}\lambda,$$

and $\boldsymbol{I}_{\tilde{\boldsymbol{\varepsilon}}} = 1/(2\pi n)\{\sum_{t=1}^{n} \tilde{\boldsymbol{\varepsilon}}(t)\mathrm{e}^{\mathrm{i}t\lambda}\}\{\sum_{t=1}^{n} \tilde{\boldsymbol{\varepsilon}}(t)\mathrm{e}^{\mathrm{i}t\lambda}\}^*$.

Assumption 8.6 (i) Θ is a compact subset of $\mathbb{R}^k$ where the dimension k is known, and $\boldsymbol{\theta}_1 \neq \boldsymbol{\theta}_2 \in \Theta$ implies $\boldsymbol{f}_{\boldsymbol{\theta}_1} \neq \boldsymbol{f}_{\boldsymbol{\theta}_2}$ on a set of positive Lebesgue measure;
(ii) $\boldsymbol{f}_{\boldsymbol{\theta}}(\lambda)$ is positive-definite, and every component of $\boldsymbol{f}_{\boldsymbol{\theta}}(\lambda)$ is also continuous in $\boldsymbol{\theta}$ and λ for $\boldsymbol{\theta} \in \Theta$ and $\lambda \in [-\pi, \pi]$;
(iii) $\arg\min_{\boldsymbol{\theta}\in\Theta} D(\boldsymbol{f}, \boldsymbol{\theta})$ is unique;
(iv) $\boldsymbol{f}(\lambda) \in Lip(\alpha)$, the Lipschitz class of degree $\alpha > 1/2$.

□

Assumption 8.7

$$\sum_{s=-(n-1)}^{n-1} \frac{\partial}{\partial \theta} \alpha_s(\boldsymbol{\theta}) = o(\sqrt{n}),$$

where $\alpha_s = 1/(2\pi) \int_{-\pi}^{\pi} \mathrm{e}^{\mathrm{i}\lambda s} \mathrm{tr}\{\boldsymbol{f}_{\boldsymbol{\theta}}(\lambda)^{-1}\}\, \mathrm{d}\lambda$. □

Lemma 8.1 *Suppose that Assumptions 8.1–8.7 hold, then, as $n \to \infty$, $\hat{\boldsymbol{\theta}} \to_p \boldsymbol{\theta}_0$, and $\sqrt{n}(\hat{\boldsymbol{\theta}} - \boldsymbol{\theta}_0) \to_d N(0, \boldsymbol{V})$, where $\boldsymbol{\theta}_0$ is the true value of $\boldsymbol{\theta}$ (i.e., $\boldsymbol{f}(\lambda) = \boldsymbol{f}_{\boldsymbol{\theta}_0}(\lambda)$), $\boldsymbol{V} = \boldsymbol{M}_f^{-1} \tilde{\boldsymbol{V}} \boldsymbol{M}_f^{-1}$,*

$$\boldsymbol{M}_f = \int_{-\pi}^{\pi} \left[\frac{\partial^2}{\partial \boldsymbol{\theta} \partial \boldsymbol{\theta}^\top} \mathrm{tr}\{\boldsymbol{f}_{\boldsymbol{\theta}}(\lambda)^{-1} \boldsymbol{f}(\lambda)\} + \frac{\partial^2}{\partial \boldsymbol{\theta} \partial \boldsymbol{\theta}^\top} \log \det \boldsymbol{f}_{\boldsymbol{\theta}}(\lambda) \right]_{\boldsymbol{\theta}=\boldsymbol{\theta}_0} \mathrm{d}\lambda,$$

$$\tilde{\boldsymbol{V}} = \{\tilde{V}_{jl}\} \text{ with } \tilde{V}_{jl} = 4\pi \int_{-\pi}^{\pi} \mathrm{tr} \left[\boldsymbol{f}(\lambda) \frac{\partial}{\partial \theta_j} \{\boldsymbol{f}_{\boldsymbol{\theta}}(\lambda)\}^{-1} \boldsymbol{f}(\lambda)) \frac{\partial}{\partial \theta_l} \{\boldsymbol{f}_{\boldsymbol{\theta}}(\lambda)\}^{-1} \right]_{\boldsymbol{\theta}=\boldsymbol{\theta}_0} \mathrm{d}\lambda.$$

Proof. From Lemma 3.1 of Hosoya and Taniguchi (1982), $\arg\min_{\boldsymbol{\theta}\in\Theta} D(\boldsymbol{f}, \boldsymbol{\theta}) = \boldsymbol{\theta}_0$, implying that

$$\frac{\partial}{\partial \theta} D(\boldsymbol{f}, \boldsymbol{\theta})|_{\boldsymbol{\theta}=\boldsymbol{\theta}_0} = 0.$$

Thus, $\boldsymbol{\theta}^*$ between $\boldsymbol{\theta}$ and $\hat{\boldsymbol{\theta}}$ exists such that

$$\frac{\partial}{\partial \theta} D(\boldsymbol{f}, \boldsymbol{\theta})|_{\boldsymbol{\theta}=\hat{\boldsymbol{\theta}}} = (\hat{\boldsymbol{\theta}} - \boldsymbol{\theta}_0) \frac{\partial^2}{\partial \boldsymbol{\theta} \partial \boldsymbol{\theta}} D(\boldsymbol{f}, \boldsymbol{\theta})|_{\boldsymbol{\theta}=\boldsymbol{\theta}^*}.$$

Furthermore, since $\hat{\boldsymbol{\theta}}$ is the minimizing value of $D(\boldsymbol{I}_{\tilde{\boldsymbol{\varepsilon}}}, \boldsymbol{\theta})$,

$$\frac{\partial}{\partial \boldsymbol{\theta}} D(\boldsymbol{I}_{\tilde{\boldsymbol{\varepsilon}}}, \boldsymbol{\theta})|_{\boldsymbol{\theta}=\hat{\boldsymbol{\theta}}} = 0.$$

Therefore,

$$\hat{\boldsymbol{\theta}} - \boldsymbol{\theta}_0 = -(\boldsymbol{M}_f^{-1} + a_n) \int_{-\pi}^{\pi} \frac{\partial}{\partial \boldsymbol{\theta}} \left(\mathrm{tr}\left[\{\boldsymbol{f}_{\boldsymbol{\theta}}(\lambda)\}^{-1} \cdot \{\boldsymbol{I}_{\tilde{\boldsymbol{\varepsilon}}}(\lambda) - \boldsymbol{f}(\lambda)\}\right]\right)_{\boldsymbol{\theta}=\boldsymbol{\theta}_0} \mathrm{d}\lambda, \quad (8.13)$$

where $\boldsymbol{M}_f^{-1} + a_n = \left(\int_{-\pi}^{\pi} \left[\frac{\partial^2}{\partial\boldsymbol{\theta}\partial\boldsymbol{\theta}^\top} \mathrm{tr}\{\boldsymbol{f}_{\boldsymbol{\theta}}(\lambda)^{-1}\boldsymbol{f}(\lambda)\} + \frac{\partial^2}{\partial\boldsymbol{\theta}\partial\boldsymbol{\theta}^\top} \log\det \boldsymbol{f}_{\boldsymbol{\theta}}(\lambda)\right]_{\boldsymbol{\theta}=\boldsymbol{\theta}^*} \mathrm{d}\lambda\right)^{-1}$.

Noticing that $\boldsymbol{I}_{\tilde{\boldsymbol{\varepsilon}}}(\lambda) = \sum_{s=-(n-1)}^{n-1} \tilde{\boldsymbol{C}}_s \mathrm{e}^{\mathrm{i}\lambda s}$, where $\tilde{\boldsymbol{C}}_s = \sum_{t=1}^{n-|s|} \tilde{\varepsilon}_t \tilde{\varepsilon}^*_{t+|s|}/n$, and that $\boldsymbol{X}(\boldsymbol{X}^\top\boldsymbol{X})^{-1}\boldsymbol{X}^\top = \boldsymbol{X}\boldsymbol{D}_n^{-1}(\boldsymbol{D}_n^{-1}\boldsymbol{X}^\top\boldsymbol{X}\boldsymbol{D}_n^{-1})^{-1}\boldsymbol{D}_n^{-1}\boldsymbol{X}^\top$ implies $\boldsymbol{C}_s - \tilde{\boldsymbol{C}}_s = O_p(1/n)$ uniformly with respect to s, where $\boldsymbol{C}_s = \sum_{t=1}^{n-|s|} \varepsilon_t \varepsilon^*_{t+|s|}/n$, we obtain that

$$\boldsymbol{I}_{\tilde{\boldsymbol{\varepsilon}}}(\lambda) = \sum_{s=-(n-1)}^{n-1} (\boldsymbol{C}_s + O_p(1/n))\mathrm{e}^{\mathrm{i}\lambda s} = \boldsymbol{I}_{\boldsymbol{\varepsilon}}(\lambda) + \sum_{s=-(n-1)}^{n-1} O_p(1/n)\mathrm{e}^{\mathrm{i}\lambda s}.$$

Thus, (8.13) is equivalent to

$$\hat{\boldsymbol{\theta}} - \boldsymbol{\theta}_0 = -(\boldsymbol{M}_f^{-1} + a_n)$$
$$\int_{-\pi}^{\pi} \frac{\partial}{\partial \boldsymbol{\theta}} \left(\mathrm{tr}\left[\{\boldsymbol{f}_{\boldsymbol{\theta}}(\lambda)\}^{-1} \cdot \{\boldsymbol{I}_{\boldsymbol{\varepsilon}}(\lambda) + \sum_{s=-(n-1)}^{n-1} O_p(1/n)\mathrm{e}^{\mathrm{i}\lambda s} - \boldsymbol{f}(\lambda)\}\right]\right)_{\boldsymbol{\theta}=\boldsymbol{\theta}_0} \mathrm{d}\lambda.$$

From Assumption 8.6(iv), following Hosoya and Taniguchi (1982), $\boldsymbol{I}_{\boldsymbol{\varepsilon}}(\lambda) \to_w \boldsymbol{f}(\lambda)$. That is

$$\int_{-\pi}^{\pi} \frac{\partial}{\partial \boldsymbol{\theta}} \left(\mathrm{tr}\left[\{\boldsymbol{f}_{\boldsymbol{\theta}}(\lambda)\}^{-1} \cdot \{\boldsymbol{I}_{\boldsymbol{\varepsilon}}(\lambda) - \boldsymbol{f}(\lambda)\}\right]\right)_{\boldsymbol{\theta}=\boldsymbol{\theta}_0} \mathrm{d}\lambda \to 0. \quad (8.14)$$

Furthermore, from Assumption 8.7,

$$\int_{-\pi}^{\pi} \frac{\partial}{\partial \boldsymbol{\theta}} \left(\mathrm{tr}\left[\{\boldsymbol{f}_{\boldsymbol{\theta}}(\lambda)\}^{-1} \sum_{s=-(n-1)}^{n-1} O_p(1/n)\mathrm{e}^{\mathrm{i}\lambda s}\right]\right)_{\boldsymbol{\theta}=\boldsymbol{\theta}_0} \mathrm{d}\lambda = o_p(1/\sqrt{n}). \quad (8.15)$$

Furthermore, (8.14) and (8.15) imply that $\hat{\boldsymbol{\theta}} \to_p \boldsymbol{\theta}_0$, and $\sqrt{n}(\hat{\boldsymbol{\theta}} - \boldsymbol{\theta}_0) \to_d N(0, \boldsymbol{V})$. □

Similar to Lemma 3.2 in Toyooka (1986), we have the following lemma.

Lemma 8.2 *Under Assumptions 8.1, 8.2 and 8.3, as $n \to \infty$, $\sqrt{n}(\hat{\boldsymbol{\theta}} - \boldsymbol{\theta}_0)$ is orthogonal with $(\boldsymbol{I}_p \otimes \boldsymbol{D}_n)^{-1}\boldsymbol{U}^\top\boldsymbol{\Gamma}^{-1}(\boldsymbol{\theta})\boldsymbol{u}$ and $(\boldsymbol{I}_p \otimes \boldsymbol{D}_n)^{-1}\boldsymbol{U}^\top\frac{\partial}{\partial\boldsymbol{\theta}}\boldsymbol{\Gamma}^{-1}(\boldsymbol{\theta})\boldsymbol{u}$, and*

$$\begin{bmatrix} (\boldsymbol{I}_p \otimes \boldsymbol{D}_n)^{-1}\boldsymbol{U}^\top\boldsymbol{\Gamma}^{-1}(\boldsymbol{\theta})\boldsymbol{u} \\ (\boldsymbol{I}_p \otimes \boldsymbol{D}_n)^{-1}\boldsymbol{U}^\top\frac{\partial}{\partial\boldsymbol{\theta}}\boldsymbol{\Gamma}^{-1}(\boldsymbol{\theta})\boldsymbol{u} \end{bmatrix} \to_d$$
$$N\left[\boldsymbol{0}, \begin{Bmatrix} \lim(\boldsymbol{I}_p \otimes \boldsymbol{D}_n)^{-1}\boldsymbol{U}^\top\boldsymbol{\Gamma}^{-1}(\boldsymbol{\theta})\boldsymbol{U}(\boldsymbol{I}_p \otimes \boldsymbol{D}_n)^{-1}, & \lim(\boldsymbol{I}_p \otimes \boldsymbol{D}_n)^{-1}\boldsymbol{U}^\top\frac{\partial}{\partial\boldsymbol{\theta}^\top}\boldsymbol{\Gamma}^{-1}(\boldsymbol{\theta})\boldsymbol{U}(\boldsymbol{I}_p \otimes \boldsymbol{D}_n)^{-1} \\ \lim(\boldsymbol{I}_p \otimes \boldsymbol{D}_n)^{-1}\boldsymbol{U}^\top\frac{\partial}{\partial\boldsymbol{\theta}}\boldsymbol{\Gamma}^{-1}(\boldsymbol{\theta})\boldsymbol{U}(\boldsymbol{I}_p \otimes \boldsymbol{D}_n)^{-1}, & \boldsymbol{\Omega} \end{Bmatrix}\right]$$

where $\boldsymbol{\Omega} = \{\Omega_{ij}\}_{k\times k}$ *and* $\Omega_{ij} = \lim(\boldsymbol{I}_p \otimes \boldsymbol{D}_n)^{-1}\boldsymbol{U}^\top \frac{\partial}{\partial\theta_i}\boldsymbol{\Gamma}^{-1}(\boldsymbol{\theta})\boldsymbol{\Gamma}(\boldsymbol{\theta})\frac{\partial}{\partial\theta_j}\boldsymbol{\Gamma}^{-1}(\boldsymbol{\theta})\boldsymbol{U}$ $(\boldsymbol{I}_p \otimes \boldsymbol{D}_n)^{-1}$.

Assumption 8.8 Set $\boldsymbol{Z} = \boldsymbol{\Gamma}^{-1}(\boldsymbol{\theta})\boldsymbol{U}$, where $\boldsymbol{Z} = \{z_1, \dots, z_q\}$, and we assume the following:

(i) $\lim_{n\to\infty} b_i^2(n) = \infty, i \in \{1, \dots, q\}$, when $b_i^2(n) = \|z_i\|^2$;

(ii) $\lim_{n\to\infty} \dfrac{\{z_i(n)\}^2}{b_i(n)^2} = 0, i \in \{1, \dots, q\}$;

(iii) For every $i, j \in \{1, \dots, q\}$ there exists the limit

$$q_{ij}(h) \equiv \lim_{n\to\infty} \frac{\sum_{t=1}^n z_i(t)z_j(t+h)}{d_i(n)d_j(n)};$$

(iv) Let $\boldsymbol{Q}_{\boldsymbol{\theta}}(h) \equiv (q_{ij}(h))$. $\boldsymbol{Q}_{\boldsymbol{\theta}}(0)$ is nonsingular. Then we can write

$$\boldsymbol{Q}_{\boldsymbol{\theta}}(h) = \int_{-\pi}^{\pi} e^{ih\lambda} \boldsymbol{N}_{\boldsymbol{\theta}}(d\lambda),$$

where $\boldsymbol{N}_{\boldsymbol{\theta}}(\lambda)$ is a matrix function whose increments are Hermitian non-negative.

□

Theorem 8.3 *Under Assumptions 8.1–8.8,*

$$\mathrm{Cov}(\boldsymbol{I}_p \otimes \boldsymbol{D}_n(\tilde{\boldsymbol{\beta}}_{BLUE})) - \mathrm{Cov}(\boldsymbol{I}_p \otimes \boldsymbol{D}_n(\hat{\boldsymbol{\beta}}_{BLUE})) = n^{-1}\sum_{i=1}^{k}\sum_{j=1}^{k} V_{ij}\boldsymbol{W}_{ij} + o(n^{-1}), \tag{8.16}$$

where V_{ij} *is the* i*th row and* j*th column element of* $\boldsymbol{V}$, *and*

$$\begin{aligned}\boldsymbol{W}_{ij} = &\lim_{n\to\infty}\{(\boldsymbol{I}_p \otimes \boldsymbol{D}_n)^{-1}\boldsymbol{U}^\top\boldsymbol{\Gamma}^{-1}(\boldsymbol{\theta})\boldsymbol{U}(\boldsymbol{I}_p \otimes \boldsymbol{D}_n)^{-1}\}^{-1}\\ &\times\left\{\lim_{n\to\infty}(\boldsymbol{I}_p \otimes \boldsymbol{D}_n)^{-1}\boldsymbol{U}^\top\frac{\partial}{\partial\theta_i}\boldsymbol{\Gamma}^{-1}(\boldsymbol{\theta})\boldsymbol{\Gamma}(\boldsymbol{\theta})\frac{\partial}{\partial\theta_j}\boldsymbol{\Gamma}^{-1}(\boldsymbol{\theta})\boldsymbol{U}(\boldsymbol{I}_p \otimes \boldsymbol{D}_n)^{-1}\right.\\ &-\lim_{n\to\infty}(\boldsymbol{I}_p \otimes \boldsymbol{D}_n)^{-1}\boldsymbol{U}^\top\frac{\partial}{\partial\theta_i}\boldsymbol{\Gamma}^{-1}(\boldsymbol{\theta})\boldsymbol{U}(\boldsymbol{I}_p \otimes \boldsymbol{D}_n)^{-1}\lim_{n\to\infty}\{(\boldsymbol{I}_p \otimes \boldsymbol{D}_n)^{-1}\boldsymbol{U}^\top\boldsymbol{\Gamma}^{-1}(\theta)\boldsymbol{U}(\boldsymbol{I}_p \otimes \boldsymbol{D}_n)^{-1}\}^{-1}\\ &\times\left.\lim_{n\to\infty}(\boldsymbol{I}_p \otimes \boldsymbol{D}_n)^{-1}\boldsymbol{U}^\top\frac{\partial}{\partial\theta_j}\boldsymbol{\Gamma}^{-1}(\boldsymbol{\theta})\boldsymbol{U}(\boldsymbol{I}_p \otimes \boldsymbol{D}_n)^{-1}\right\}\lim_{n\to\infty}\{(\boldsymbol{I}_p \otimes \boldsymbol{D}_n)^{-1}\boldsymbol{U}^\top\boldsymbol{\Gamma}^{-1}(\theta)\boldsymbol{U}(\boldsymbol{I}_p \otimes \boldsymbol{D}_n)^{-1}\}^{-1}.\end{aligned}$$

In particular, when $p = 1$,

$$\boldsymbol{W}_{ij} = \{\int \boldsymbol{f}_{\boldsymbol{\theta}}(\lambda)^{-1}\,\mathrm{d}\boldsymbol{M}(\lambda)\}^{-1}\Bigg[\int \boldsymbol{f}_{\boldsymbol{\theta}}(\lambda)^{-1}\frac{\partial}{\partial\theta_i}\boldsymbol{f}_{\boldsymbol{\theta}}(\lambda)\boldsymbol{f}_{\boldsymbol{\theta}}(\lambda)^{-1}\frac{\partial}{\partial\theta_j}\boldsymbol{f}_{\boldsymbol{\theta}}(\lambda)\boldsymbol{f}_{\boldsymbol{\theta}}(\lambda)^{-1}\,\mathrm{d}\boldsymbol{M}(\lambda)$$
$$-\frac{1}{2}\int \boldsymbol{f}_{\boldsymbol{\theta}}(\lambda)^{-1}\frac{\partial^2}{\partial\theta_i\partial\theta_j}\boldsymbol{f}_{\boldsymbol{\theta}}(\lambda)\boldsymbol{f}_{\boldsymbol{\theta}}(\lambda)^{-1}\,\mathrm{d}\boldsymbol{M}(\lambda) + \frac{1}{2}\int \frac{\partial^2}{\partial\theta_i\partial\theta_j}\boldsymbol{f}_{\boldsymbol{\theta}}(\lambda)\,\mathrm{d}\boldsymbol{N}_{\boldsymbol{\theta}}(\lambda)$$
$$-\int \boldsymbol{f}_{\boldsymbol{\theta}}(\lambda)^{-1}\frac{\partial}{\partial\theta_i}\boldsymbol{f}_{\boldsymbol{\theta}}(\lambda)\boldsymbol{f}_{\boldsymbol{\theta}}(\lambda)^{-1}\,\mathrm{d}\boldsymbol{M}(\lambda)\left\{\int \boldsymbol{f}_{\boldsymbol{\theta}}(\lambda)^{-1}\,\mathrm{d}\boldsymbol{M}(\lambda)\right\}^{-1}$$
$$\int \boldsymbol{f}_{\boldsymbol{\theta}}(\lambda)^{-1}\frac{\partial}{\partial\theta_j}\boldsymbol{f}_{\boldsymbol{\theta}}(\lambda)\boldsymbol{f}_{\boldsymbol{\theta}}(\lambda)^{-1}\,\mathrm{d}\boldsymbol{M}(\lambda)\Bigg]\left\{\int \boldsymbol{f}_{\boldsymbol{\theta}}(\lambda)^{-1}\,\mathrm{d}\boldsymbol{M}(\lambda)\right\}^{-1}.$$

Proof. Following the line of Toyooka (1986), we have

$$\mathrm{Cov}(\boldsymbol{I}_p \otimes \boldsymbol{D}_n(\tilde{\boldsymbol{\beta}}_{\mathrm{BLUE}} - \boldsymbol{\beta})) = \mathrm{Cov}(\boldsymbol{I}_p \otimes \boldsymbol{D}_n\hat{\boldsymbol{\beta}}_{\mathrm{BLUE}})$$
$$+\,\mathrm{e}\left[\boldsymbol{I}_p \otimes \boldsymbol{D}_n(\tilde{\boldsymbol{\beta}}_{\mathrm{BLUE}} - \hat{\boldsymbol{\beta}}_{\mathrm{BLUE}})(\tilde{\boldsymbol{\beta}}_{\mathrm{BLUE}} - \hat{\boldsymbol{\beta}}_{\mathrm{BLUE}})^\top \boldsymbol{I}_p \otimes \boldsymbol{D}_n\right].$$

Notice that

$$\mathrm{E}\left[\boldsymbol{I}_p \otimes \boldsymbol{D}_n(\tilde{\boldsymbol{\beta}}_{\mathrm{BLUE}} - \hat{\boldsymbol{\beta}}_{\mathrm{BLUE}})(\tilde{\boldsymbol{\beta}}_{\mathrm{BLUE}} - \hat{\boldsymbol{\beta}}_{\mathrm{BLUE}})^\top \boldsymbol{I}_p \otimes \boldsymbol{D}_n\right]$$
$$= \mathrm{E}\left[\boldsymbol{I}_p \otimes \boldsymbol{D}_n\left\{(\tilde{\boldsymbol{\beta}}_{\mathrm{BLUE}} - \boldsymbol{\beta}) - (\hat{\boldsymbol{\beta}}_{\mathrm{BLUE}} - \boldsymbol{\beta})\right\}\right.$$
$$\left.\left\{(\tilde{\boldsymbol{\beta}}_{\mathrm{BLUE}} - \boldsymbol{\beta}) - (\hat{\boldsymbol{\beta}}_{\mathrm{BLUE}} - \boldsymbol{\beta})\right\}^\top \boldsymbol{I}_p \otimes \boldsymbol{D}_n\right],$$

$$\tilde{\boldsymbol{\beta}}_{\mathrm{BLUE}} - \boldsymbol{\beta} = (\boldsymbol{U}^\top\boldsymbol{\Gamma}^{-1}(\hat{\boldsymbol{\theta}})\boldsymbol{U})^{-1}\boldsymbol{U}^\top\boldsymbol{\Gamma}^{-1}(\hat{\boldsymbol{\theta}})\boldsymbol{u}, \qquad \hat{\boldsymbol{\beta}}_{\mathrm{BLUE}} - \boldsymbol{\beta} = (\boldsymbol{U}^\top\boldsymbol{\Gamma}^{-1}(\boldsymbol{\theta})\boldsymbol{U})^{-1}\boldsymbol{U}^\top\boldsymbol{\Gamma}^{-1}(\boldsymbol{\theta})\boldsymbol{u}.$$

From Lemmas 8.1 and 8.2, we have

$$\mathrm{E}\left[\boldsymbol{I}_p \otimes \boldsymbol{D}_n(\tilde{\boldsymbol{\beta}}_{\mathrm{BLUE}} - \hat{\boldsymbol{\beta}}_{\mathrm{BLUE}})(\tilde{\boldsymbol{\beta}}_{\mathrm{BLUE}} - \hat{\boldsymbol{\beta}}_{\mathrm{BLUE}})^\top \boldsymbol{I}_p \otimes \boldsymbol{D}_n\right] =$$
$$\sum_{i=1}^{k}\sum_{j=1}^{k} n^{-1}V_{ij}\lim_{n\to\infty}\{(\boldsymbol{I}_p \otimes \boldsymbol{D}_n)^{-1}\boldsymbol{U}^\top\boldsymbol{\Gamma}^{-1}(\boldsymbol{\theta})\boldsymbol{U}(\boldsymbol{I}_p \otimes \boldsymbol{D}_n)^{-1}\}^{-1}$$
$$\Bigg\{\lim_{n\to\infty}(\boldsymbol{I}_p \otimes \boldsymbol{D}_n)^{-1}\boldsymbol{U}^\top\frac{\partial}{\partial\theta_i}\boldsymbol{\Gamma}^{-1}(\boldsymbol{\theta})\boldsymbol{\Gamma}(\boldsymbol{\theta})\frac{\partial}{\partial\theta_j}\boldsymbol{\Gamma}^{-1}(\boldsymbol{\theta})\boldsymbol{U}(\boldsymbol{I}_p \otimes \boldsymbol{D}_n)^{-1}$$
$$-\lim_{n\to\infty}(\boldsymbol{I}_p \otimes \boldsymbol{D}_n)^{-1}\boldsymbol{U}^\top\frac{\partial}{\partial\theta_i}\boldsymbol{\Gamma}^{-1}(\boldsymbol{\theta})\boldsymbol{U}(\boldsymbol{I}_p \otimes \boldsymbol{D}_n)^{-1}\lim_{n\to\infty}\{(\boldsymbol{I}_p \otimes \boldsymbol{D}_n)^{-1}\boldsymbol{U}^\top\boldsymbol{\Gamma}^{-1}(\theta)\boldsymbol{U}(\boldsymbol{I}_p \otimes \boldsymbol{D}_n)^{-1}\}^{-1}$$
$$\times\lim_{n\to\infty}(\boldsymbol{I}_p \otimes \boldsymbol{D}_n)^{-1}\boldsymbol{U}^\top\frac{\partial}{\partial\theta_j}\boldsymbol{\Gamma}^{-1}(\boldsymbol{\theta})\boldsymbol{U}(\boldsymbol{I}_p \otimes \boldsymbol{D}_n)^{-1}\Bigg\}$$
$$\times\lim_{n\to\infty}\{(\boldsymbol{I}_p \otimes \boldsymbol{D}_n)^{-1}\boldsymbol{U}^\top\boldsymbol{\Gamma}^{-1}(\boldsymbol{\theta})\boldsymbol{U}(\boldsymbol{I}_p \otimes \boldsymbol{D}_n)^{-1}\}^{-1} + o(n^{-1}) = n^{-1}\sum_{i=1}^{k}\sum_{j=1}^{k}V_{ij}\boldsymbol{W}_{ij} + o(n^{-1}).$$

Furthermore, we remark that

$$\frac{\partial}{\partial\theta_i}\boldsymbol{\Gamma}^{-1}(\boldsymbol{\theta}) = -\boldsymbol{\Gamma}^{-1}(\boldsymbol{\theta})\frac{\partial}{\partial\theta_i}\boldsymbol{\Gamma}(\boldsymbol{\theta})\boldsymbol{\Gamma}^{-1}(\boldsymbol{\theta}),$$

and

$$\frac{\partial^2}{\partial\theta_i\partial\theta_j}\boldsymbol{\Gamma}^{-1}(\boldsymbol{\theta})=\boldsymbol{\Gamma}^{-1}(\boldsymbol{\theta})\frac{\partial}{\partial\theta_i}\boldsymbol{\Gamma}(\boldsymbol{\theta})\boldsymbol{\Gamma}^{-1}(\boldsymbol{\theta})\frac{\partial}{\partial\theta_j}\boldsymbol{\Gamma}(\boldsymbol{\theta})\boldsymbol{\Gamma}^{-1}(\boldsymbol{\theta})$$
$$+\boldsymbol{\Gamma}^{-1}(\boldsymbol{\theta})\frac{\partial}{\partial\theta_j}\boldsymbol{\Gamma}(\boldsymbol{\theta})\boldsymbol{\Gamma}^{-1}(\boldsymbol{\theta})\frac{\partial}{\partial\theta_i}\boldsymbol{\Gamma}(\boldsymbol{\theta})\boldsymbol{\Gamma}^{-1}(\boldsymbol{\theta})-\boldsymbol{\Gamma}^{-1}(\boldsymbol{\theta})\frac{\partial^2}{\partial\theta_i\partial\theta_j}\boldsymbol{\Gamma}(\boldsymbol{\theta})\boldsymbol{\Gamma}^{-1}(\boldsymbol{\theta}).$$

Thus,

$$\begin{aligned}
&\mathrm{e}\left[\boldsymbol{I}_p\otimes\boldsymbol{D}_n(\tilde{\boldsymbol{\beta}}_{\mathrm{BLUE}}-\hat{\boldsymbol{\beta}}_{\mathrm{BLUE}})(\tilde{\boldsymbol{\beta}}_{\mathrm{BLUE}}-\hat{\boldsymbol{\beta}}_{\mathrm{BLUE}})^\top\boldsymbol{I}_p\otimes\boldsymbol{D}_n\right]\\
&=\sum_{i=1,j=1}^{i=k,j=k}n^{-1}V_{ij}\lim_{n\to\infty}\{(\boldsymbol{I}_p\otimes\boldsymbol{D}_n)^{-1}\boldsymbol{U}^\top\boldsymbol{\Gamma}^{-1}(\boldsymbol{\theta})\boldsymbol{U}(\boldsymbol{I}_p\otimes\boldsymbol{D}_n)^{-1}\}^{-1}\\
&\quad\times\Bigg\{\lim_{n\to\infty}(\boldsymbol{I}_p\otimes\boldsymbol{D}_n)^{-1}\boldsymbol{U}^\top\frac{1}{2}\left[\frac{\partial^2}{\partial\theta_i\partial\theta_j}\boldsymbol{\Gamma}^{-1}(\boldsymbol{\theta})+\boldsymbol{\Gamma}^{-1}(\boldsymbol{\theta})\frac{\partial^2}{\partial\theta_i\partial\theta_j}\boldsymbol{\Gamma}(\boldsymbol{\theta})\boldsymbol{\Gamma}^{-1}(\boldsymbol{\theta})\right]\boldsymbol{U}(\boldsymbol{I}_p\otimes\boldsymbol{D}_n)^{-1}\\
&\quad-\lim_{n\to\infty}(\boldsymbol{I}_p\otimes\boldsymbol{D}_n)^{-1}\boldsymbol{U}^\top\frac{\partial}{\partial\theta_i}\boldsymbol{\Gamma}^{-1}(\boldsymbol{\theta})\boldsymbol{U}(\boldsymbol{I}_p\otimes\boldsymbol{D}_n)^{-1}\lim_{n\to\infty}\{(\boldsymbol{I}_p\otimes\boldsymbol{D}_n)^{-1}\boldsymbol{U}^\top\boldsymbol{\Gamma}^{-1}(\theta)\boldsymbol{U}(\boldsymbol{I}_p\otimes\boldsymbol{D}_n)^{-1}\}^{-1}\\
&\qquad\times\lim_{n\to\infty}(\boldsymbol{I}_p\otimes\boldsymbol{D}_n)^{-1}\boldsymbol{U}^\top\frac{\partial}{\partial\theta_j}\boldsymbol{\Gamma}^{-1}(\boldsymbol{\theta})\boldsymbol{U}(\boldsymbol{I}_p\otimes\boldsymbol{D}_n)^{-1}\Bigg\}\\
&\qquad\lim_{n\to\infty}\{(\boldsymbol{I}_p\otimes\boldsymbol{D}_n)^{-1}\boldsymbol{U}^\top\boldsymbol{\Gamma}^{-1}(\boldsymbol{\theta})\boldsymbol{U}(\boldsymbol{I}_p\otimes\boldsymbol{D}_n)^{-1}\}^{-1}+o(n^{-1}).
\end{aligned}$$

When $p=1$, from Assumption 8.8, we have

$$\begin{aligned}
&\mathrm{e}\left[\boldsymbol{I}_p\otimes\boldsymbol{D}_n(\tilde{\boldsymbol{\beta}}_{\mathrm{BLUE}}-\hat{\boldsymbol{\beta}}_{\mathrm{BLUE}})(\tilde{\boldsymbol{\beta}}_{\mathrm{BLUE}}-\hat{\boldsymbol{\beta}}_{\mathrm{BLUE}})^\top\boldsymbol{I}_p\otimes\boldsymbol{D}_n\right]=\\
&n^{-1}\sum_{i=1,j=1}^{i=k,j=k}V_{ij}\left\{\int\boldsymbol{f}_{\boldsymbol{\theta}}(\lambda)^{-1}\,\mathrm{d}\boldsymbol{M}(\lambda)\right\}^{-1}\times\Bigg[\int\boldsymbol{f}_{\boldsymbol{\theta}}(\lambda)^{-1}\frac{\partial}{\partial\theta_i}\boldsymbol{f}_{\boldsymbol{\theta}}(\lambda)\boldsymbol{f}_{\boldsymbol{\theta}}(\lambda)^{-1}\frac{\partial}{\partial\theta_j}\boldsymbol{f}_{\boldsymbol{\theta}}(\lambda)\boldsymbol{f}_{\boldsymbol{\theta}}(\lambda)^{-1}\,\mathrm{d}\boldsymbol{M}(\lambda)\\
&\quad-\frac{1}{2}\int\boldsymbol{f}_{\boldsymbol{\theta}}(\lambda)^{-1}\frac{\partial^2}{\partial\theta_i\partial\theta_j}\boldsymbol{f}_{\boldsymbol{\theta}}(\lambda)\boldsymbol{f}_{\boldsymbol{\theta}}(\lambda)^{-1}\,\mathrm{d}\boldsymbol{M}(\lambda)+\frac{1}{2}\int\frac{\partial^2}{\partial\theta_i\partial\theta_j}\boldsymbol{f}_{\boldsymbol{\theta}}(\lambda)\,\mathrm{d}\boldsymbol{N}_{\boldsymbol{\theta}}(\lambda)\\
&\quad-\int\boldsymbol{f}_{\boldsymbol{\theta}}(\lambda)^{-1}\frac{\partial}{\partial\theta_i}\boldsymbol{f}_{\boldsymbol{\theta}}(\lambda)\boldsymbol{f}_{\boldsymbol{\theta}}(\lambda)^{-1}\,\mathrm{d}\boldsymbol{M}(\lambda)\times\left\{\int\boldsymbol{f}_{\boldsymbol{\theta}}(\lambda)^{-1}\,\mathrm{d}\boldsymbol{M}(\lambda)\right\}^{-1}\\
&\quad\times\int\boldsymbol{f}_{\boldsymbol{\theta}}(\lambda)^{-1}\frac{\partial}{\partial\theta_j}\boldsymbol{f}_{\boldsymbol{\theta}}(\lambda)\boldsymbol{f}_{\boldsymbol{\theta}}(\lambda)^{-1}\,\mathrm{d}\boldsymbol{M}(\lambda)\Bigg]\times\left\{\int\boldsymbol{f}_{\boldsymbol{\theta}}(\lambda)^{-1}\,\mathrm{d}\boldsymbol{M}(\lambda)\right\}^{-1}+o(n^{-1}).
\end{aligned}\tag{8.17}$$

□

Corollary 8.1 *Under Assumptions 8.1–8.8, when* $p=1$*, if* $\boldsymbol{M}(\lambda)$ *and* $\boldsymbol{N}_{\boldsymbol{\theta}}(\lambda)$ *each have only one jump at* $\lambda=0$*, where the jumps are* $\boldsymbol{R}$ *and* $\boldsymbol{f}_{\boldsymbol{\theta}}(0)^{-2}\boldsymbol{R}$*, respectively, then*

$$\mathrm{Cov}(\boldsymbol{I}_p\otimes\boldsymbol{D}_n(\tilde{\boldsymbol{\beta}}_{BLUE}))-\mathrm{Cov}(\boldsymbol{I}_p\otimes\boldsymbol{D}_n(\hat{\boldsymbol{\beta}}_{BLUE}))=o(n^{-1}).$$

Proof. Noticing that $\boldsymbol{M}(\lambda)$ and $\boldsymbol{N}_{\boldsymbol{\theta}}(\lambda)$ each have only one jump at $\lambda=0$ with jumps $\boldsymbol{R}$ and $\boldsymbol{f}_{\boldsymbol{\theta}}(0)^{-2}\boldsymbol{R}$, respectively, in (8.17), the coefficient of n^{-1} vanishes. Hence, the Corollary holds. □

For $\tilde{\beta}$, we have

$$\begin{aligned}\mathrm{DMSE}_n(\hat{\boldsymbol{\beta}}_{\mathrm{BLUE}}, \tilde{\boldsymbol{\beta}}) &\equiv \mathrm{MSE}_n(\hat{\boldsymbol{\beta}}_{\mathrm{BLUE}}) - \mathrm{MSE}_n(\tilde{\boldsymbol{\beta}}) = \mathrm{MSE}_n(\tilde{\boldsymbol{\beta}}_{\mathrm{BLUE}}) - \mathrm{MSE}_n(\tilde{\boldsymbol{\beta}}) - n^{-1}\mathrm{tr}\sum_{i=1}^{k}\sum_{j=1}^{k} V_{ij}W_{ij} \\ &= -c^2 \mathrm{e}\left[\frac{1}{\|\boldsymbol{I}_p \otimes \boldsymbol{D}_n \tilde{\boldsymbol{\beta}}_{\mathrm{BLUE}}\|^2}\right] + 2c\left(1 - \mathrm{e}\left[\frac{< \boldsymbol{I}_p \otimes \boldsymbol{D}_n \boldsymbol{\beta}, \boldsymbol{I}_p \otimes \boldsymbol{D}_n \tilde{\boldsymbol{\beta}}_{\mathrm{BLUE}} >}{\|\boldsymbol{I}_p \otimes \boldsymbol{D}_n \tilde{\boldsymbol{\beta}}_{\mathrm{BLUE}}\|^2}\right]\right) \\ &\quad - n^{-1}\mathrm{tr}\sum_{i=1}^{k}\sum_{j=1}^{k} V_{ij}W_{ij} - o(n^{-1}) \end{aligned} \tag{8.18}$$

Theorem 8.4 *In the case of $\boldsymbol{\beta} = \mathbf{0}$, suppose that Assumptions 8.1–8.8 hold and that $pq \geq 3$. Then,*

(i)

$$c\left\{2 - \frac{c}{pq-2}\left(\frac{\hat{\nu}_{pq,n}}{\hat{\nu}_{1,n}}\right)^{pq/2}\frac{1}{\hat{\nu}_{p,n}}\right\} \leq \mathrm{DMSE}_n(\hat{\boldsymbol{\beta}}_{\mathrm{BLUE}}, \tilde{\boldsymbol{\beta}}) \leq c\left\{2 - \frac{c}{pq-2}\left(\frac{\hat{\nu}_{1,n}}{\hat{\nu}_{pq,n}}\right)^{pq/2}\frac{1}{\hat{\nu}_{1,n}}\right\}, \tag{8.19}$$

which implies that $\tilde{\boldsymbol{\beta}}$ improves $\hat{\boldsymbol{\beta}}_{\mathrm{BLUE}}$ if the LHS of (8.19) is positive;

(ii)

$$c\left\{2 - \frac{c}{pq-2}\left(\frac{\nu_{pq}}{\nu_1}\right)^{pq/2}\frac{1}{\nu_{pq}}\right\} \leq \lim_{n\to\infty}\mathrm{DMSE}_n(\hat{\boldsymbol{\beta}}_{\mathrm{BLUE}}, \tilde{\boldsymbol{\beta}}) \leq c\left\{2 - \frac{c}{pq-2}\left(\frac{\nu_1}{\nu_{pq}}\right)^{pq/2}\frac{1}{\nu_1}\right\}, \tag{8.20}$$

which implies that $\tilde{\boldsymbol{\beta}}$ improves $\hat{\boldsymbol{\beta}}_{\mathrm{BLUE}}$ asymptotically if the LHS of (8.20) is positive,

where $\hat{\nu}_{1,n}, \ldots, \hat{\nu}_{pq,n}$ ($\hat{\nu}_{1,n} \leq \cdots \leq \hat{\nu}_{pq,n}$) and $\nu_1, \ldots, \nu_{pq}$ ($\nu_1 \leq \cdots \leq \nu_{pq}$) are the eigenvalues of $\hat{\boldsymbol{C}}_n$ and $\boldsymbol{C}$, respectively, and $\hat{\boldsymbol{C}}_n = (\boldsymbol{I}_p \otimes \boldsymbol{D}_n)(\boldsymbol{U}^\top\boldsymbol{\Gamma}^{-1}(\hat{\boldsymbol{\theta}})\boldsymbol{U})^{-1}\boldsymbol{U}^\top\boldsymbol{\Gamma}^{-1}(\hat{\boldsymbol{\theta}})\boldsymbol{U}(\boldsymbol{U}^\top\boldsymbol{\Gamma}^{-1}(\hat{\boldsymbol{\theta}})\boldsymbol{U})^{-1}(\boldsymbol{I}_p \otimes \boldsymbol{D}_n)$.

In the case of $\boldsymbol{\beta} \neq \mathbf{0}$,

$$\mathrm{DMSE}_n(\hat{\boldsymbol{\beta}}_{\mathrm{BLUE}}, \tilde{\boldsymbol{\beta}}) = \frac{2c}{\|(\boldsymbol{I}_p \otimes \boldsymbol{D}_n)\boldsymbol{\beta}\|^2}\{\Delta_n(c, \boldsymbol{\beta}) + o(1)\} - n^{-1}\mathrm{tr}\sum_{i=1}^{k}\sum_{j=1}^{k} V_{ij}W_{ij} - o(n^{-1}), \tag{8.21}$$

where $\Delta_n(c, \boldsymbol{\beta}) = \mathrm{tr}\boldsymbol{C}_n - 2\boldsymbol{\beta}^\top(\boldsymbol{I}_p \otimes \boldsymbol{D}_n)\boldsymbol{C}_n(\boldsymbol{I}_p \otimes \boldsymbol{D}_n)\boldsymbol{\beta}/\|(\boldsymbol{I}_p \otimes \boldsymbol{D}_n)\boldsymbol{\beta}\|^2 - c/2$. If $\kappa > 0$, by noticing that $O(\|(\boldsymbol{I}_p \otimes \boldsymbol{D}_n)\boldsymbol{\beta}\|^2) > O(n)$, $\mathrm{DMSE}_n(\hat{\boldsymbol{\beta}}_{\mathrm{BLUE}}, \tilde{\boldsymbol{\beta}})$ tends to be smaller than 0. Here we consider the situation that $\kappa = 0$. Typically, we remark that under the conditions of Corollary 8.1, the coefficient of n^{-1} in (8.18) vanishes, which implies the discussion of DMSE is much the same as that in Sect. 8.3. We omit the discussion.

Theorem 8.5 *In the case of $\boldsymbol{\beta} = \boldsymbol{0}$, suppose that Assumptions 8.1–8.8 hold and that $pq \geq 3$. Then, if $c > 0$*

(i) we have the inequality

$$\Delta_n(c, \boldsymbol{\beta}) \geq \sum_{j=1}^{pq-1} \hat{\nu}_{j,n} - \hat{\nu}_{pq,n} - \frac{c}{2}, \tag{8.22}$$

which implies that $\tilde{\boldsymbol{\beta}}$ tends to improve $\hat{\boldsymbol{\beta}}_{\mathrm{BLUE}}$ up to n^{-1}-order if

$$\frac{2c}{\|(\boldsymbol{I}_p \otimes \boldsymbol{D}_n)\boldsymbol{\beta}\|^2}\Big(\sum_{j=1}^{pq-1} \hat{\nu}_{j,n} - \hat{\nu}_{pq,n} - \frac{c}{2}\Big) - n^{-1}\mathrm{tr}\sum_{i=1}^{k}\sum_{j=1}^{k} V_{ij} W_{ij},$$

is positive.

(ii) Taking the limit of (8.22), we have

$$\lim_{n\to\infty} \Delta_n(c, \boldsymbol{\beta}) \geq \sum_{j=1}^{pq-1} \nu_j - \nu_{pq} - \frac{c}{2},$$

which implies that $\tilde{\boldsymbol{\beta}}$ tends to improve $\hat{\boldsymbol{\beta}}_{\mathrm{BLUE}}$ asymptotically up to n^{-1}-order if

$$\lim_{n\to\infty} \frac{2cn}{\|(\boldsymbol{I}_p \otimes \boldsymbol{D}_n)\boldsymbol{\beta}\|^2}\Big(\sum_{j=1}^{pq-1} \nu_j - \nu_{pq} - \frac{c}{2}\Big) - \mathrm{tr}\sum_{i=1}^{k}\sum_{j=1}^{k} V_{ij} W_{ij},$$

is positive.

The proofs of Theorems 8.4 and 8.5 are similar to Theorems 8.1 and 8.2, we omit these here.

8.5 Example and Numerical Analysis

In this section, we compare the MSEs of LSE $\hat{\boldsymbol{\beta}}_{\mathrm{LS}}$, James-Stein estimator $\hat{\boldsymbol{\beta}}_{\mathrm{JS}}$, BLUE $\hat{\boldsymbol{\beta}}_{\mathrm{BLUE}}$, the shrinkage estimator based on BLUE $\hat{\boldsymbol{\beta}}$, the feasible estimators $\tilde{\boldsymbol{\beta}}_{\mathrm{BLUE}}$

Table 8.1 The settings of coefficients of $\{\varepsilon(t)\}$ and the corresponding model and spectral density for each scenario defined in

	θ_1	θ_2	model	f
Scenario 1	0.1	0	AR(1)	$\frac{0.1}{2\pi}/\|1+0.1e^{i\lambda}\|^2$
Scenario 2	0.9	0	AR(1)	$\frac{0.1}{2\pi}/\|1+0.9e^{i\lambda}\|^2$
Scenario 3	0	0.1	MA(1)	$\frac{0.1}{2\pi}\|1+0.1e^{i\lambda}\|^2$
Scenario 4	0	0.9	MA(1)	$\frac{0.1}{2\pi}\|1+0.9e^{i\lambda}\|^2$

and $\tilde{\boldsymbol{\beta}}$ when $\boldsymbol{\beta} = \mathbf{0}$ and $\boldsymbol{\beta} \neq \mathbf{0}$ corresponding to the theorems in Sects. 8.3 and 8.4. Specifically, we set $\boldsymbol{b} = (0, \ldots, 0)^\top$ for $\hat{\boldsymbol{\beta}}_{\mathrm{JS}}$. For the linear regression model:

$$\boldsymbol{y}(t) = \boldsymbol{B}^\top \boldsymbol{x}(t) + \boldsymbol{\varepsilon}(t), \qquad t \in \{1, \ldots, n\}, \tag{8.23}$$

we set $p = 1, q = 8$, and $x_i(t) = \cos \lambda_i t, i \in \{1, \ldots, q\}$ with $\lambda_i = \pi(i-1)/9$. Thus $\boldsymbol{\beta} = \boldsymbol{B}$, $\boldsymbol{U} = \boldsymbol{X}$, and $\boldsymbol{I}_p \otimes \boldsymbol{D}_n = \boldsymbol{D}_n$. For $\boldsymbol{X}$, we obtain that $\kappa = 0$, $d_i(n)^2 = O(n)$ and from Chapter 10 in Anderson (1971)

$$\rho_{ij}(h) = \begin{cases} \cos \lambda_i h, & i = j, \\ 0, & i \neq j, \end{cases}$$

and $\boldsymbol{M}(\lambda)$ has a jump $\boldsymbol{M}_j = \mathrm{diag}\{0, \ldots, 0, 1/2, 0, \ldots, 0\}$ at $\lambda = \pm\lambda_j$ where $1/2$ is at the jth diagonal element, implying that Assumptions 8.1–8.4 hold. Suppose that the spectral density of $\{\varepsilon(t)\}$ is $f(\lambda)$, $\boldsymbol{C} = \mathrm{diag}\{2\pi f(\lambda_1), \ldots, 2\pi f(\lambda_q)\}$. To generate the error sequence $\{\varepsilon(t)\}$, we draw $\{\varepsilon(t)\}$ using following model:

$$\varepsilon(t) + \theta_1 \varepsilon(t-1) = \eta(t) + \theta_2 \eta(t-1),$$

where $\{\eta(t)\}$ is an i.i.d. sequence from a $N(0, 0.1)$ distribution; the settings of the coefficients are listed in Table 8.1 to guarantee Assumptions 8.5 and 8.6 hold. Noticing that Scenarios 1 and 2 correspond to AR(1) models, $\frac{\partial}{\partial\theta}\alpha_s(\boldsymbol{\theta})$s is the Fourier coefficients of the first derivative of a MA(1) model's spectral density function with θ located in $(-1, 1)$. Thus

$$\lim_{n\to\infty} \sum_{s=-(n-1)}^{n-1} \frac{\partial}{\partial\theta}\alpha_s(\boldsymbol{\theta}) = \frac{\partial}{\partial\theta} f_\theta(0),$$

which implies that Assumption 8.7 holds. Regarding Assumption 8.8, from the Gershgorin circle theorem, the eigenvalues of $\Gamma(\theta)$ are bounded in $[0.1/\{2\pi(1+\theta^2+2|\theta|)\}, 0.1(1+2|\theta|/(1-\theta)/\{2\pi(1-\theta^2)\}]$ (say $[B_1, B_2]$). Hence $B_1 d_i(n) \leq b_i(n) \leq B_2 d_i(n)$, and then Assumption 8.8 holds. Similarly, for Scenarios 3 and 4, Assumptions 8.7 and 8.8 also hold.

For the settings of $\boldsymbol{\beta}$, we consider three cases: $\boldsymbol{\beta} = (0, \ldots, 0)^\top$, $\boldsymbol{\beta} = (0.1, \ldots, 0.1)^\top$, and $\boldsymbol{\beta} = (1, \ldots, 1)^\top$. For the cs in the form of shrinkage estimators, it is natural that the greater DMSE is, the more suitably c is chosen. From the theorems introduced in Sect. 8.4, we give the lower bound of DMSE_n under certain conditions. The c where the lower bound obtains its maximum value is optimal theoretically. Hence, we take c as the value where the lower bound of DMSE_n gets the maximum corresponding to the above cases. For example, when $\boldsymbol{\beta} = (0, \ldots, 0)^\top$, c is taken as the maximum point of the LHS of (8.6) to estimate $\hat{\boldsymbol{\beta}}$, and similarly c is taken as the maximum point on the LHS of (8.19) to estimate $\tilde{\boldsymbol{\beta}}$. Specifically, when $\boldsymbol{\beta} \neq (0, \ldots, 0)^\top$, for AR(1) and MA(1), the dimension of θ is $k = 1$, and since $p = 1$, $n^{-1} \sum_{i=1}^{k} \sum_{j=1}^{k} V_{ij} \boldsymbol{W}_{ij}$ in (8.18) can be simplified by

$$
\begin{aligned}
n^{-1} \sum_{i=1}^{k} \sum_{j=1}^{k} V_{ij} \boldsymbol{W}_{ij} &= n^{-1} V \boldsymbol{W} = n^{-1} V \left\{ \int \boldsymbol{f}_{\theta}(\lambda)^{-1} \, \mathrm{d}\boldsymbol{M}(\lambda) \right\}^{-1} \\
&\times \left[\int \boldsymbol{f}_{\theta}(\lambda)^{-1} \frac{\partial}{\partial \theta} \boldsymbol{f}_{\theta}(\lambda) \boldsymbol{f}_{\theta}(\lambda)^{-1} \frac{\partial}{\partial \theta} \boldsymbol{f}_{\theta}(\lambda) \boldsymbol{f}_{\theta}(\lambda)^{-1} \, \mathrm{d}\boldsymbol{M}(\lambda) \right. \\
&- \frac{1}{2} \int \boldsymbol{f}_{\theta}(\lambda)^{-1} \frac{\partial^2}{\partial \theta^2} \boldsymbol{f}_{\theta}(\lambda) \boldsymbol{f}_{\theta}(\lambda)^{-1} \, \mathrm{d}\boldsymbol{M}(\lambda) + \frac{1}{2} \int \frac{\partial^2}{\partial \theta^2} \boldsymbol{f}_{\theta}(\lambda) \, \mathrm{d}\boldsymbol{N}_{\theta}(\lambda) \\
&- \int \boldsymbol{f}_{\theta}(\lambda)^{-1} \frac{\partial}{\partial \theta} \boldsymbol{f}_{\theta}(\lambda) \boldsymbol{f}_{\theta}(\lambda)^{-1} \, \mathrm{d}\boldsymbol{M}(\lambda) \times \left\{ \int \boldsymbol{f}_{\theta}(\lambda)^{-1} \, \mathrm{d}\boldsymbol{M}(\lambda) \right\}^{-1} \\
&\left. \times \int \boldsymbol{f}_{\theta}(\lambda)^{-1} \frac{\partial}{\partial \theta} \boldsymbol{f}_{\theta}(\lambda) \boldsymbol{f}_{\theta}(\lambda)^{-1} \, \mathrm{d}\boldsymbol{M}(\lambda) \right] \times \left\{ \int \boldsymbol{f}_{\theta}(\lambda)^{-1} \, \mathrm{d}\boldsymbol{M}(\lambda) \right\}^{-1}.
\end{aligned}
$$

Noticing that $M(\lambda)$ has jumps $\boldsymbol{M}_j = \mathrm{diag}\{0, \ldots, 0, 1/2, 0, \ldots, 0\}$ at $\lambda = \pm \lambda_j$ where $1/2$ is the jth diagonal element, we have

$$
\begin{aligned}
n^{-1} V \boldsymbol{W} &= n^{-1} V \left\{ \sum_{i=1}^{q} \frac{1}{2} (\boldsymbol{f}_{\theta}(\lambda_i)^{-1} + \boldsymbol{f}_{\theta}(-\lambda_i)^{-1}) \right\}^{-1} \\
&\times \left[\frac{1}{2} \sum_{i=1}^{q} \boldsymbol{f}_{\theta}(\lambda_i)^{-1} \frac{\partial}{\partial \theta} \boldsymbol{f}_{\theta}(\lambda_i) \boldsymbol{f}_{\theta}(\lambda_i)^{-1} \frac{\partial}{\partial \theta} \boldsymbol{f}_{\theta}(\lambda_i) \boldsymbol{f}_{\theta}(\lambda_i)^{-1} \right. \\
&+ \frac{1}{2} \sum_{i=1}^{q} \boldsymbol{f}_{\theta}(-\lambda_i)^{-1} \frac{\partial}{\partial \theta} \boldsymbol{f}_{\theta}(-\lambda_i) \boldsymbol{f}_{\theta}(-\lambda_i)^{-1} \frac{\partial}{\partial \theta} \boldsymbol{f}_{\theta}(-\lambda_i) \boldsymbol{f}_{\theta}(-\lambda_i)^{-1} \\
&- \frac{1}{4} \sum_{i=1}^{q} \boldsymbol{f}_{\theta}(\lambda_i)^{-1} \frac{\partial^2}{\partial \theta^2} \boldsymbol{f}_{\theta}(\lambda_i) \boldsymbol{f}_{\theta}(\lambda_i)^{-1} \quad - \frac{1}{4} \sum_{i=1}^{q} \boldsymbol{f}_{\theta}(-\lambda_i)^{-1} \frac{\partial^2}{\partial \theta^2} \boldsymbol{f}_{\theta}(-\lambda_i) \boldsymbol{f}_{\theta}(-\lambda_i)^{-1} \\
&+ \frac{1}{4} \sum_{i=1}^{q} \boldsymbol{f}_{\theta}(\lambda_i)^{-1} \frac{\partial^2}{\partial \theta^2} \boldsymbol{f}_{\theta}(\lambda_i) \boldsymbol{f}_{\theta}(\lambda_i)^{-1} + \frac{1}{4} \sum_{i=1}^{q} \boldsymbol{f}_{\theta}(-\lambda_i)^{-1} \frac{\partial^2}{\partial \theta^2} \boldsymbol{f}_{\theta}(-\lambda_i) \boldsymbol{f}_{\theta}(-\lambda_i)^{-1} \\
&- \left(\frac{1}{2} \sum_{i=1}^{q} \boldsymbol{f}_{\theta}(\lambda_i)^{-1} \frac{\partial}{\partial \theta} \boldsymbol{f}_{\theta}(\lambda_i) \boldsymbol{f}_{\theta}(\lambda_i)^{-1} + \frac{1}{2} \sum_{i=1}^{q} \boldsymbol{f}_{\theta}(-\lambda_i)^{-1} \frac{\partial}{\partial \theta} \boldsymbol{f}_{\theta}(-\lambda_i) \boldsymbol{f}_{\theta}(-\lambda_i)^{-1} \right) \\
&\left\{ \sum_{i=1}^{q} \frac{1}{2} (\boldsymbol{f}_{\theta}(\lambda_i)^{-1} + \boldsymbol{f}_{\theta}(-\lambda_i)^{-1}) \right\}^{-1} \\
&\left. \times \left(\frac{1}{2} \sum_{i=1}^{q} \boldsymbol{f}_{\theta}(\lambda_i)^{-1} \frac{\partial}{\partial \theta} \boldsymbol{f}_{\theta}(\lambda_i) \boldsymbol{f}_{\theta}(\lambda_i)^{-1} + \frac{1}{2} \sum_{i=1}^{q} \boldsymbol{f}_{\theta}(-\lambda_i)^{-1} \frac{\partial}{\partial \theta} \boldsymbol{f}_{\theta}(-\lambda_i) \boldsymbol{f}_{\theta}(-\lambda_i)^{-1} \right) \right]
\end{aligned}
$$

$$\times \left\{ \sum_{i=1}^{q} \frac{1}{2} (f_\theta(\lambda_i)^{-1} + f_\theta(-\lambda_i)^{-1}) \right\}^{-1}.$$

Noticing that $f_\theta(\lambda_i) = f_\theta(-\lambda_i)$, $n^{-1} V W = 0$, which implies that $n^{-1} \sum_{i=1}^{k} \sum_{j=1}^{k} V_{ij} W_{ij}$ can be ignored in (8.18) when choosing c. In Table 8.2, we calculate the MSE of each estimator by using Monte Carlo simulations with sample size 100 and 100 times replications, where the MSEs are approximated by

$$\begin{aligned} \mathrm{MSE}(\boldsymbol{\phi}(\boldsymbol{X}, \boldsymbol{Y})) = & \frac{1}{100-1} \sum_{i=1}^{100} \| \boldsymbol{D}_n \{ \boldsymbol{\phi}(\boldsymbol{X}, \boldsymbol{Y})_i - \frac{1}{100} \sum_{j=1}^{100} \boldsymbol{\phi}(\boldsymbol{X}, \boldsymbol{Y})_j \} \|^2 \\ & + \| \boldsymbol{D}_n \{ \frac{1}{100} \sum_{j=1}^{100} \boldsymbol{\phi}(\boldsymbol{X}, \boldsymbol{Y})_j - \boldsymbol{\beta} \} \|^2, \end{aligned}$$

where $\boldsymbol{\phi}(\boldsymbol{X}, \boldsymbol{Y})$ is an arbitrary estimator of $\boldsymbol{\beta}$.

It is shown that both the MSEs of $\hat{\boldsymbol{\beta}}$ and $\tilde{\boldsymbol{\beta}}$ are smaller than those of $\hat{\boldsymbol{\beta}}_{\mathrm{BLUE}}$ and $\tilde{\boldsymbol{\beta}}_{\mathrm{BLUE}}$. This result agrees with the previous theorems, and it implies that the feasible estimator $\tilde{\boldsymbol{\beta}}$ is effective. In addition, it indicates that when c is taken as the maximum point of DMSE_n, the shrinkage estimators based on BLUE perform better than the James-Stein estimator which is a type of shrinkage estimator based on LSE when the dependence among the error sequence is strong. It implies that for strong dependent error sequence, we may prefer $\tilde{\boldsymbol{\beta}}$ rather than $\hat{\boldsymbol{\beta}}_{\mathrm{JS}}$ to estimate the coefficients. Furthermore, we notice that the improvement is becoming larger when $\boldsymbol{\beta}$ is close to $\boldsymbol{\beta} = (0, \ldots, 0)^\top$, which means that if the difference between $\boldsymbol{\beta}$ and $\boldsymbol{b}$ is small, the improvement may be notable. In addition, we examine the MSEs of $\hat{\boldsymbol{\beta}}$ with respect to different c to check whether the theoretically optimal c performs better than the other c. c is taken from $(c_{\min} + 0/10 * (c_{\max} - c_{\min}), c_{\min} + 1/10 * (c_{\max} - c_{\min}), \ldots, c_{\min} + 10/10 * (c_{\max} - c_{\min}))$ where the lower bound of DMSE_n is positive in the interval $(c_{\min}, c_{\max})$ and obtains its maximum value at point $c_{\min} + 5/10 * (c_{\max} - c_{\min})$ (i.e., the theoretically optimal c), and $\{\varepsilon(t)\}$ follows AR(1) with $\theta_1 = 0.1$ and MA(1) with $\theta_2 = 0.1$, respectively. In Figs. 8.1 and 8.2, the yellow line represents the MSEs of BLUE estimator, the blue line is the MSEs of $\hat{\boldsymbol{\beta}}$ with respect to c, and the vertical black line represents the theoretically optimal c. From these two figures, we can obtain that DMSE_n does not get the maximum value at the optimal point. However, at this point, DMSE_n is relatively sufficiently large. In Fig. 8.2c, part of the blue line is over the yellow line which implies that DMSE_n is negative. This is because that for the case $\beta \neq 0$, there is an infinitesimal $o(1)$ in (8.8), whose influence cannot be ignored when the sample size is not sufficiently large. Furthermore, from Fig. 8.1, and Fig. 8.2a and b, we can know that c may exist beyond the interval where (8.6) and (8.9) hold, such that $\hat{\boldsymbol{\beta}}$ improves $\hat{\boldsymbol{\beta}}_{\mathrm{BLUE}}$ as well. This is for the reason that (8.6) and (8.9) are sufficient but not necessary conditions where $\hat{\boldsymbol{\beta}}$ improves $\hat{\boldsymbol{\beta}}_{\mathrm{BLUE}}$. For cs where (8.6) and (8.9) do not hold, the possibility of improvement exists. We should note that when $\boldsymbol{\beta} \neq \boldsymbol{0}$, if $\sum_{j=1}^{pq-1} \nu_{j,n} - \nu_{pq,n}$ is

Table 8.2 The MSEs of β_{LS}, $\hat{\boldsymbol{\beta}}_{\mathrm{JS}}$, $\hat{\boldsymbol{\beta}}_{\mathrm{BLUE}}$, $\hat{\boldsymbol{\beta}}$, $\tilde{\boldsymbol{\beta}}_{\mathrm{BLUE}}$ and $\tilde{\boldsymbol{\beta}}$ in the cases of $\boldsymbol{\beta} = (0, \ldots, 0)^\top$, $\boldsymbol{\beta} = (0.1, \ldots, 0.1)^\top$, and $\boldsymbol{\beta} = (1, \ldots, 1)^\top$

MSE				
	AR(1), $\theta_1 = 0.1$	AR(1), $\theta_1 = 0.9$	MA(1), $\theta_2 = 0.1$	MA(1), $\theta_2 = 0.9$
$\boldsymbol{\beta} = (0, \ldots, 0)^\top$				
β_{LS}	0.7624	0.7298	0.8548	0.9783
$\hat{\boldsymbol{\beta}}_{\mathrm{JS}}$	0.4619	0.7293	0.5640	0.9783
$\hat{\boldsymbol{\beta}}_{\mathrm{BLUE}}$	0.7621	0.6548	0.8534	0.9638
$\hat{\boldsymbol{\beta}}$	0.4617	0.6543	0.5628	0.9638
$\tilde{\boldsymbol{\beta}}_{\mathrm{BLUE}}$	0.7622	0.6538	0.8526	0.9647
$\tilde{\boldsymbol{\beta}}$	0.4628	0.6531	0.5988	0.9647
$\boldsymbol{\beta} = (0.1, \ldots, 0.1)^\top$				
β_{LS}	0.7624	0.7298	0.8548	0.9783
$\hat{\boldsymbol{\beta}}_{\mathrm{JS}}$	0.6771	0.7069	0.7799	0.8785
$\hat{\boldsymbol{\beta}}_{\mathrm{BLUE}}$	0.7621	0.6548	0.8534	0.9638
$\hat{\boldsymbol{\beta}}$	0.6767	0.6327	0.7790	0.8647
$\tilde{\boldsymbol{\beta}}_{\mathrm{BLUE}}$	0.7625	0.6538	0.8526	0.9647
$\tilde{\boldsymbol{\beta}}$	0.6774	0.6305	0.7800	0.8667
$\boldsymbol{\beta} = (1, \ldots, 1)^\top$				
β_{LS}	0.7624	0.7298	0.8548	0.9783
$\hat{\boldsymbol{\beta}}_{\mathrm{JS}}$	0.7593	0.7295	0.8542	0.9766
$\hat{\boldsymbol{\beta}}_{\mathrm{BLUE}}$	0.7621	0.6548	0.8534	0.9638
$\hat{\boldsymbol{\beta}}$	0.7591	0.6544	0.8529	0.9621
$\tilde{\boldsymbol{\beta}}_{\mathrm{BLUE}}$	0.7625	0.6538	0.8526	0.9647
$\tilde{\boldsymbol{\beta}}$	0.7595	0.6533	0.8522	0.9630

negative, we cannot find a positive c for $\hat{\boldsymbol{\beta}}$, which guarantees the positivity of the lower bound of DMSE_n, and similarly, there is not a suitable positive c for $\tilde{\boldsymbol{\beta}}$ if $\sum_{j=1}^{pq-1} \hat{\nu}_{j,n} - \hat{\nu}_{pq,n}$ is negative. In the above model (8.23), if we reset $\boldsymbol{x}(t)$ as $x_i(t) = \cos \lambda_i t$, $i \in \{1, \ldots, q\}$ with $\lambda_i = \pi(i+1)/9$, for $\theta_1 = 0.9$ and $\theta_2 = 0$, we cannot find suitable cs for $\hat{\boldsymbol{\beta}}$ and $\tilde{\boldsymbol{\beta}}$. Table 8.3 shows the comparison of MSEs when c is negative and is the maximizing value of the lower bound of DMSE_n. The MSEs of shrinkage estimators are not better than their corresponding estimators. The result is consistent with the fact that the lower bound does not make sense with negative c. In such situation, $\tilde{\boldsymbol{\beta}}_{\mathrm{BLUE}}$ performs better than the others when solving the practical problems.

Table 8.3 The MSEs of β_{LS}, $\hat{\boldsymbol{\beta}}_{\text{JS}}$, $\hat{\boldsymbol{\beta}}_{\text{BLUE}}$, $\hat{\boldsymbol{\beta}}$, $\tilde{\boldsymbol{\beta}}_{\text{BLUE}}$, and $\tilde{\boldsymbol{\beta}}$ when the "optimal" cs for $\hat{\boldsymbol{\beta}}_{\text{JS}}$, $\hat{\boldsymbol{\beta}}$ and $\tilde{\boldsymbol{\beta}}$ are negative

MSE		
	$\boldsymbol{\beta} = (0.1, \ldots, 0.1)^\top$	$\boldsymbol{\beta} = (1, \ldots, 1)^\top$
β_{LS}	1.4526	1.4526
$\hat{\boldsymbol{\beta}}_{\text{JS}}$	1.5906	1.4527
$\hat{\boldsymbol{\beta}}_{\text{BLUE}}$	1.3447	1.3447
$\hat{\boldsymbol{\beta}}$	1.4780	1.3449
$\tilde{\boldsymbol{\beta}}_{\text{BLUE}}$	1.3464	1.3464
$\tilde{\boldsymbol{\beta}}$	1.4602	1.3465

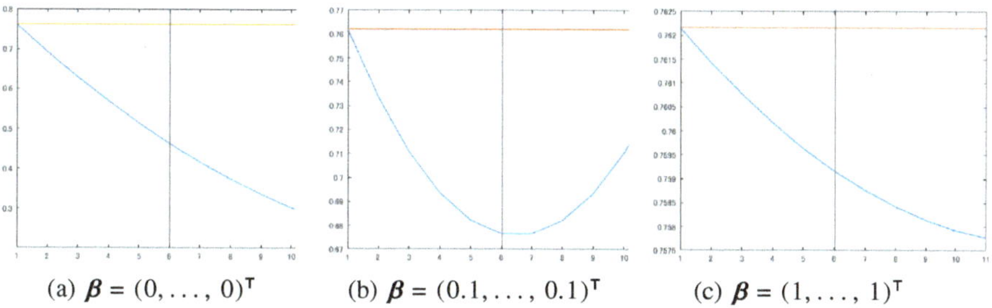

(a) $\boldsymbol{\beta} = (0, \ldots, 0)^\top$ (b) $\boldsymbol{\beta} = (0.1, \ldots, 0.1)^\top$ (c) $\boldsymbol{\beta} = (1, \ldots, 1)^\top$

Fig. 8.1 The curves of MSEs of $\hat{\boldsymbol{\beta}}_{\text{BLUE}}$ (yellow line) and $\hat{\boldsymbol{\beta}}$ (blue line) with respect to c taken from $(c_{\min} + 0/10 * (c_{\max} - c_{\min}), c_{\min} + 1/10 * (c_{\max} - c_{\min}), \ldots, c_{\min} + 10/10 * (c_{\max} - c_{\min}))$ where the lower bound of DMSE$_n$ is positive in the interval $(c_{\min}, c_{\max})$ and obtains its maximum value at point $c_{\min} + 5/10 * (c_{\max} - c_{\min})$, and $\{\varepsilon(t)\}$ follows AR(1) with $\theta_1 = 0.1$

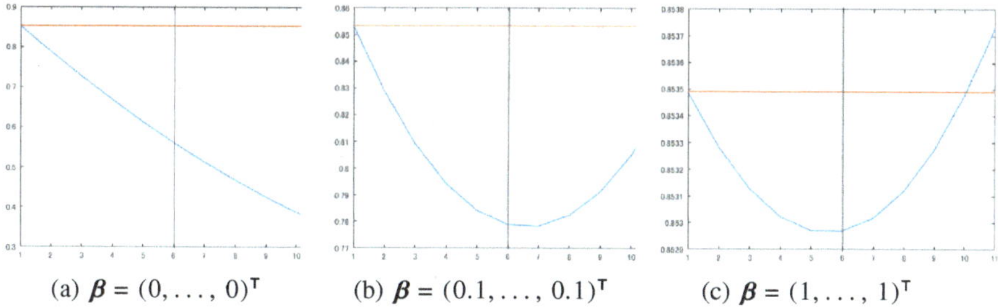

(a) $\boldsymbol{\beta} = (0, \ldots, 0)^\top$ (b) $\boldsymbol{\beta} = (0.1, \ldots, 0.1)^\top$ (c) $\boldsymbol{\beta} = (1, \ldots, 1)^\top$

Fig. 8.2 The curves of MSEs of $\hat{\boldsymbol{\beta}}_{\text{BLUE}}$ (yellow line) and $\hat{\boldsymbol{\beta}}$ (blue line) with respect to c taken from $(c_{\min} + 0/10 * (c_{\max} - c_{\min}), c_{\min} + 1/10 * (c_{\max} - c_{\min}), \ldots, c_{\min} + 10/10 * (c_{\max} - c_{\min}))$ where the lower bound of DMSE$_n$ is positive in the interval $(c_{\min}, c_{\max})$ and obtains its maximum value at point $c_{\min} + 5/10 * (c_{\max} - c_{\min})$, and $\{\varepsilon(t)\}$ follows MA(1) with $\theta_2 = 0.1$

8.6 Conclusions

We propose a shrinkage estimator based on BLUE which has a better performance than LSE in the sense of MSE when the residual process is dependent, and the sufficient conditions when the shrinkage estimator improves BLUE with respect to MSE is provided. Furthermore, since the autocovariance matrix of the residual process is infeasible, we introduce a feasible approximator of that shrinkage estimator by replacing the autocovariance matrix by an estimator of it, where the approach is extended from Toyooka (1986). Correspondingly, we also present the sufficient conditions where the feasible version improves BLUE. Furthermore, we also present results of some numerical studies. From these results, it is shown that the feasible version of that shrinkage estimator is effective, and that for strong dependent residual process, it may be better to prefer it rather than $\hat{\boldsymbol{\beta}}_{\mathrm{JS}}$ which is a kind of shrinkage estimator based on LSE to estimate the coefficients of linear models.

References

Anderson, Theodore W (1971). *The statistical analysis of time series*. New York: Springer.

Arnold, S. F. (1981). *The Theory of Linear Models and Multivariate Analysis*. New York: Wiley.

Efron, B. and C. Morris (1975). "Data analysis using Stein's estimator and its generalizations". In: *J. Am. Stat. Assoc* 70.350, pp. 311–319.

Grenander, U. and M. Rosenblatt (1957). *Statistical Analysis of Stationary Time Series*. New York: John Wiley & Sons.

Hannan, E. J. (1970). *Multiple Time Series*. New York: Wiley.

He, Shuyuan. (1995). "Uniform convergency for weighted periodogram of stationary linear random fields". In: *Chin. Ann. of Math.* 3, pp. 331–340.

Hosoya, Y. and M. Taniguchi (1982). "A central limit theorem for stationary process and the parameter estimation of linear processes". In: *Ann. Stat.* 10.1, pp. 132–153.

James, W. and C. Stein (1961). "Estimation with quadratic loss". In: *Proc. Fourth Berkeley Symp. Math. Statist. Prob.* 1, pp. 361–380.

Lahiri, P. and J. N. K. Rao (1995). "Robust estimation of mean squared error of small area estimators". In: *JASA* 90.430, pp. 758–766.

Senda, M. and M. Taniguchi (2006). "James-Stein estimators for time series regression models". In: *J. Multivariate Anal.* 97, pp. 1984–1996.

Shiraishi, H., M. Taniguchi, and T. Yamashita (2018). "Higher-order asymptotic theory of shrinkage estimation for general statistical models". In: *J. Multivariate Anal.* 166, pp. 198–211.

Stein, C. (1956). "Inadmissibility of the Usual Estimator for the Mean of a Multivariate Normal Distribution". In: *Proc. Third Berkeley Symp. Math. Statist. Prob.* 1, pp. 197–206.

Taniguchi, M. (1991). *Higher Order Asymptotic Theory for Multivariate Time Series*. New York: Springer.

Taniguchi, M. and J. Hirukawa (2005). "The Stein-James estimator for short- and long-memory Gaussian processes". In: *Biometrika* 92, pp. 737–746.

Taniguchi, M. and Y. Kakizawa (2020). *Asymptotic Theory of Statistical Inference for Time Series*. New York: Springer.

Toyooka, Y. (1986). "Second-order risk structure of GLSE and MLE in a regression with a linear process". In: *Ann. Statist.* 14, pp. 1214–1225.

Xue, Yujie, Masanobu Taniguchi, and Tong Liu (2024). "Shrinkage estimators of BLUE for time series regression models". In: *Journal of Multivariate Analysis* 202, p. 105282.

Zeitfracht Medien GmbH
Ferdinand-Jühlke-Straße 7
99095 Erfurt, Deutschland
produktsicherheit@kolibri360.de